"Hill has written a very accesitive benefits of accepting evolution theory and Christian faith as compatible. What impresses me most, though, is the intellectually humble way in which he addresses the topic. Unlike much of what we see from our political and religious leaders today, Hill does not claim that his perspective alone is valid and that those who differ either lack the requisite knowledge or have questionable motives. Hill acknowledges the challenges facing honest, thoughtful Christians who think seriously about evolution and their faith and invites them to consider—rather than trying to coerce them to accept—the ways in which faith and evolution can be brought into profitable integration."

David Basinger, professor of philosophy and ethics at Roberts Wesleyan College, coauthor of *Reason and Religious Belief*

"Some question whether biblical faith is compatible with the findings of natural science, especially regarding human origins. Matthew Hill engages that question with genuine respect. Even those untroubled by evolutionary biology may be surprised and encouraged by his further claim that our evolutionary nature has significance for our growth in Christlikeness. Combining sound knowledge of the issues and a down-to-earth style, *Embracing Evolution* invites thoughtful readers and engaged study groups."

Joel B. Green, professor of New Testament interpretation at Fuller Theological Seminary

"In recent years there has been a notable increase in books outlining ways to fully embrace both Christian faith and modern science. Scholars from a variety of academic disciplines in the sciences and humanities have provided fruitful contributions in understanding this complex topic. Dr. Hill's book is a valuable addition to this discussion from the perspective of an ordained pastor who is also a professional philosopher. His provocative thesis—understanding science can strengthen your Christian life—is one that I have experienced both personally as a follower of Jesus and academically as a biologist and theologian. Hill's book is well written and accessible to a wide audience. His many stories are captivating and complement his scholarship. This is a book I would encourage all Christians to add to their library."

Denis O. Lamoureux, professor of science and religion at St. Joseph's College, University of Alberta

EMBRACING EVOLUTION

HOW UNDERSTANDING SCIENCE CAN STRENGTHEN YOUR CHRISTIAN LIFE

MATTHEW NELSON HILL

FOREWORD BY J. RICHARD MIDDLETON

An imprint of InterVarsity Press
Downers Grove, Illinois

InterVarsity Press
P.O. Box 1400, Downers Grove, IL 60515-1426
ivpress.com
email@ivpress.com

©2020 by Matthew Nelson Hill

All rights reserved. No part of this book may be reproduced in any form without written permission from InterVarsity Press.

InterVarsity Press® is the book-publishing division of InterVarsity Christian Fellowship/USA®, a movement of students and faculty active on campus at hundreds of universities, colleges, and schools of nursing in the United States of America, and a member movement of the International Fellowship of Evangelical Students. For information about local and regional activities, visit intervarsity.org.

Scripture quotations, unless otherwise noted, are from the New Revised Standard Version of the Bible, copyright 1989 by the Division of Christian Education of the National Council of the Churches of Christ in the USA. Used by permission. All rights reserved.

While any stories in this book are true, some names and identifying information may have been changed to protect the privacy of individuals.

Cover design and image composite: David Fassett
Interior design: Jeanna Wiggins
Images: fog over forest: © Mraust / iStockphoto
  forest and clouds: © huseyintuncer / iStockphoto
  profile of a woman: © panic_attack / iStockphoto
  Swiss Alps: © cdbrphotography / iStockphoto
  night sky: © michal-rojek / iStockphoto
  flock of pigeons: © mustafagull / iStockphoto
  zebra herd: © WLDavies / iStockphoto
  dahlia petals: © Ogphoto / iStockphoto
  DNA strand: © 3xy / iStockphoto
  flying birds: © Kent Odelli / EyeEm / Getty Images

ISBN 978-0-8308-5283-3 (print)
ISBN 978-0-8308-3923-0 (digital)

InterVarsity Press is committed to ecological stewardship and to the conservation of natural resources in all our operations. This book was printed using sustainably sourced paper.

Library of Congress Cataloging-in-Publication Data
A catalog record for this book is available from the Library of Congress.

| P | 25 | 24 | 23 | 22 | 21 | 20 | 19 | 18 | 17 | 16 | 15 | 14 | 13 | 12 | 11 | 10 | 9 | 8 | 7 | 6 | 5 | 4 | 3 | 2 | 1 |
| Y | 41 | 40 | 39 | 38 | 37 | 36 | 35 | 34 | 33 | 32 | 31 | 30 | 29 | 28 | 27 | 26 | 25 | 24 | 23 | 22 | 21 | 20 | | | | |

This book is dedicated to

Helene—with all my love—and to my children:

Connor, Anna, Lucas, and Eva.

May you always know where you come from

and what kingdom you belong to.

CONTENTS

Foreword by J. Richard Middleton	ix
Acknowledgments	xv
1 Opening a Dialogue	1

PART ONE: Understanding Our Biblical Lens

2 Reading Scripture Faithfully	15
3 Adam and Eve, the Fall, Predation, and Death	32

PART TWO: Understanding Our Scientific Lens

4 The Nuts and Bolts of Evolution	45
5 Relating to Science	70

PART THREE: An Integrated Approach to Evolution and the Christian Faith

6 Understanding Evolutionary Theory Can Be Empowering	87
7 Having Evolutionary Roots Isn't Just Baggage: Pursuing a Holistic Understanding of Redemption	102
8 Nurturing Natural Virtues Toward Healthy Community	116
Study Guide	131
Bibliography	133

FOREWORD

J. Richard Middleton

MANY CHRISTIANS TODAY are on a journey of understanding, trying to make sense of evolution in light of their faith. This is particularly difficult in our polarized cultural climate in North America, where religion and science are often portrayed as opposed to each other.

For that reason, I am delighted to be able to write this foreword to Matt Hill's *Embracing Evolution*. Whereas many books on Christian faith and evolution either view the two as antithetical to each other or struggle to make significant connections between them, *Embracing Evolution* shows that understanding human evolution can be positively helpful for Christians seeking to be faithful to Jesus Christ.

MY JOURNEY OF UNDERSTANDING THE BIBLE AND SCIENCE ON ORIGINS

Unlike those Christians who started out as young earth creationists and became convinced of the validity of biological evolution later in life, I have no memory of ever dismissing evolution as fundamentally incompatible with biblical faith. Having become a Christian at a young age, I not only accepted in my teenage years that the earth was very old (based on what seemed to be reasonable scientific research) but as a young adult read widely about the evolution of *Homo sapiens* and our various hominin relatives.

Thankfully, my home church in Kingston, Jamaica (Grace Missionary Church), never insisted on young earth creationism. And when I began my undergraduate studies at Jamaica Theological Seminary, I took two courses in my first semester that made such a view of creation untenable.

The first was a course on the Pentateuch, where one of the textbooks assigned was Bernard Ramm's *The Christian View of Science and Scripture*. Here I found an evangelical theologian outlining multiple views of how the Bible related to a variety of scientific issues. Although Ramm articulated his own opinion on the issues he discussed, he noted that there was no single obvious "biblical" answer for questions such as the age of the earth, the great flood, or even evolution. In each case, this was a matter not of biblical authority but of scientific evidence.

In my first undergraduate semester I also took a course on hermeneutics, or biblical interpretation, where the textbook was A. Berkeley Mickelsen's *Interpreting the Bible*. While this was a bit of a dense read for an eighteen-year-old, I never forgot Mickelsen's point that since there was no human observer at creation and since the eschaton is still future, biblical language describing the beginning and end must be largely figurative; these descriptions inevitably transcended human experience. Therefore, just as it would be inappropriate to read eschatological imagery in the book of Revelation as a journalistic account of what a movie camera might record (which seemed obvious to me), I came to realize that it would likewise be a misreading of Genesis to treat the six days of creation as a scientific account of origins.

These two courses at the start of my theological studies combined to convince me that there was no conflict, in principle, between science and the Bible on the question of origins. More than that, these courses (along with the rest of my seminary education) encouraged me to be open to the scientific exploration of God's world.

During my undergraduate studies I was also developing an interest in a holistic theology that affirmed the goodness of creation (in the beginning) and God's intent to redeem the cosmos (in the end).[1] By the time I graduated with my bachelor of theology degree, I was on a track to take seriously what the sciences were telling us about how this world, including biological life, came to be.

COGNITIVE DISSONANCE ABOUT EVOLUTION

Then as a graduate student in philosophy working as a campus minister for InterVarsity Christian Fellowship at the University of Guelph in Canada, I found myself avidly reading books on hominin evolution—including *Lucy*, the account of the discovery of *Australopithecus afarensis* (nicknamed Lucy) by Donald Johanson and Maitland Edey.

Although I had no real doubts about the scientific evidence for evolution, including the evolution of *Homo sapiens*, I was somewhat troubled that evolution didn't seem compatible with the biblical notion of the fall, the origin of evil recounted in Genesis 2–3. I had always been taught that this text portrays Adam and Eve (an original couple) forfeiting a primal paradisiacal state through a single act of disobedience, which led to the introduction of death for both humans and the natural world. I couldn't get my head around how this might fit with what scientists claimed about human evolution, including the obvious fact that animal and plant death preceded the origin of humanity on earth. So I did what many Christians do when confronted with cognitive dissonance—I put it out of my mind and concentrated on other things.

In my case, these other things were my graduate studies, first a master's degree in philosophy and then course work in Old Testament, fol-

---

[1]This led to a book that I coauthored with Brian J. Walsh, *The Transforming Vision: Shaping a Christian World View* (Downers Grove, IL: IVP Academic, 1984); I later wrote a book specifically on eschatology, *A New Heaven and a New Earth: Reclaiming Biblical Eschatology* (Grand Rapids: Baker Academic, 2014).

lowed by a doctoral dissertation on humans as *imago Dei* in Genesis 1 (published as *The Liberating Image: The Imago Dei in Genesis 1*).

In the years leading up to my dissertation, I taught often on the *imago Dei* in both church and academic settings, and I've now written some dozen articles and blog posts on the subject. I have also regularly taught on the garden story of Genesis 2–3, both in churches and in a variety of undergraduate and graduate courses.

My teaching on the first three chapters of Genesis was developed without any explicit reference to evolution. Rather, my focus was on how these texts should be read for their theological discernment of God, the world, and the human calling. Instead of referencing the modern scientific context, I was focused on how the theology of ancient Israel, gleaned from the Bible itself, along with the "cognitive environment" of the ancient Near East, contributed to the meaning of these texts for the life of the church.[2]

EVOLUTION AND THE FALL

But everything changed in 2013, when I was invited by James K. A. Smith to join an interdisciplinary team of scholars (united by a commitment to the classic orthodox creeds of the church) who would connect their scholarly expertise to the subject of human evolution and the fall. The invitation to participate in this project set me on a path to address the very questions that my cognitive dissonance had previously led me to avoid.

As I began working on how the narrative of Genesis 2–3 might relate to the evolutionary history of *Homo sapiens*, I discovered that paying attention to evolution did not detract from reading the text but actually helped me notice nuances that I had previously overlooked. For example, I had simply assumed that the first humans lived in a

---

[2]For more on the term *cognitive environment*, see John H. Walton, *Ancient Near Eastern Thought and the Old Testament: Introducing the Conceptual World of the Hebrew Bible*, 2nd ed. (Grand Rapids: Baker Academic, 2018). Walton has popularized this idea through his Lost World series.

paradisiacal state of perfection before the entrance of sin. Yet immediately after the creation of humanity in Genesis 2, we have the account of human disobedience in Genesis 3. Might that lack of narration of a paradisiacal state be significant for relating the text to evolutionary history?

In the essay I wrote on Genesis 2–3, published in a volume called *Evolution and the Fall*, I attempted to hold together an evolutionary account of humanity with a real historical origin of evil (which I believe is a nonnegotiable Christian doctrine), yet without claiming that the Bible and science are saying the same thing.[3]

In doing so, I was rejecting the classic idea that we can easily correlate or harmonize the Bible and science. Yet I also found Stephen Jay Gould's famous idea of nonoverlapping magisteria (NOMA) inadequate. This view is usually taken to mean that the Bible and science describe different realms of reality—and so cannot, in principle, contradict one another. However, I have now come to formulate the relationship between the Bible and science as two different lenses or perspectives through which we may view the same world.

Of course, the connections between the lenses of the garden story and human evolution aren't seamless. As Matt Hill himself admits, it isn't always easy to correlate what the Bible tells us theologically about suffering and death with the history of animal predation and extinctions long before humans came along. And how exactly does a biblical perspective on human sin relate to the development of moral consciousness among *Homo sapiens*—or even among earlier hominins?

EVOLUTION AND THE CHRISTIAN LIFE

But *Embracing Evolution* does not focus on the Bible and science generally. Instead, the book addresses how knowledge of evolution

---

[3] J. Richard Middleton, "Reading Genesis 3 Attentive to Human Evolution: Beyond Concordism and Non-Overlapping Magisteria," in *Evolution and the Fall*, ed. William T. Cavanaugh and James K. A. Smith (Grand Rapids: Eerdmans, 2017), chap. 4.

can aid us in the quest for holiness and moral transformation in the Christian life.

Matt helpfully builds on his earlier (more technical) book, *Evolution and Holiness: Sociobiology, Altruism and the Quest for Wesleyan Perfection*, but with a wider purview. Drawing on what we know about our common genetic inheritance as human beings, and even the specific proclivities we may have because of our particular ancestry, Matt gives practical advice on how this knowledge can help us make better moral decisions as we seek to be faithful to the God of the Scriptures.

Having done more and more speaking of late for church groups and conferences on how a biblical approach to questions of human identity and the origin of evil might be related to what the sciences are telling us about human evolution, I've found a hunger among Christians (and interested others) to come to a deeper understanding of biblical faith in a way that opens us up to learning from God's other book, the empirical world that the sciences address.

I am delighted to recommend Matt Hill's *Embracing Evolution* as a wonderful addition to the literature on this subject.

ACKNOWLEDGMENTS

OVER THE YEARS I'VE MET dozens of people who have turned away from their faith—unnecessarily so—because they were told they had to choose between faith and science. As believers, we often add extra epistemological requirements to people's faith that are not biblical and are otherwise disingenuous to the gospel. Having seen too many friends and acquaintances leave a God that they cannot see for scientific evidence that they can measure, I became wholly fed up with this false choice. We can certainly do better—we have to do better—for the sake of our children.

When IVP extended an invitation to me to be part of this project, I was a bit intimidated to write for a popular audience. After several drafts, false starts, wasted months, and mini-meltdowns, I reflect back on the work of this manuscript with much gratitude to the individuals and institutions that have helped me along the way.

I would like to thank David Congdon for initially asking me to write this book and for starting me on this project, and for Jon Boyd for keeping that momentum going. Many thanks to the folks who helped me brainstorm ideas for chapters, including but not limited to Brent Cline, Robbie Bolton, Cameron Moore, Barsanuphius and Ben Keaster, Ken Brewer, Robert Moore-Jumonville, Tom Holsinger-Friesen, Elisee Ouoba, and the theology department at Spring Arbor

University, as well as Michael Dowd, Denis Lamoureux, Darrel R. Falk, Peter Enns, Tom Oord, Jim Stump, Curtis Hotlzen, and Howard Snyder. And to Richard Middleton for his generous foreword and helpful comments.

I would also like to thank David and Janet McKenna and the McKenna grant that enabled me to fund this project—not to mention their endorsement and support. Thanks also to my bishops in the Free Methodist Church for their encouragement. To early readers of my drafts, Kyle Poag, Andrew Smith, Craig Welkener, Tom Corbett, and Steve Castle, I appreciate your honest critiques, criticisms, and reassurances. Thanks also to Destiny Sykes for copyedits and help with the study guide.

I am particularly grateful to Jack Baker for his tireless edits and critiques. Every suggestion he made—every hour he spent on this manuscript—made it stronger and more focused. I have learned so much from his teaching and encouragement.

I would also like to thank my parents, Nelson and Karen Hill, for their unending support. I feel so grateful for their model of how Christians should live in this world.

Last, and by far most important, I would like to thank my family: Helene, Connor, Anna, Lucas, and Eva. My wife, Helene, has shown me what real Christianity looks like with her unending faith, her relentless optimism, and her limitless altruistic spirit. I can't imagine what life would be like without that living witness to follow. And to my children: may you always ask the hard questions. And may you always remember that Christ's sacrificial and incarnate love isn't just the greatest story that has ever been, it's the greatest story that *could have ever* been. May you always find the truth in it. May you always tell others about the truth in it. And may you not be afraid of your biological roots and, somehow through the community of other believers, nurture your natural proclivities for the glory of God and for the hope of your neighbor.

1

OPENING A DIALOGUE

A FASCINATING EXPERIMENT, the Marshmallow Test, was designed by Walter Mischel of Stanford University in the 1960s. The study involved six hundred kids (four to six years old) and sought to assess levels of self-control and delayed gratification. It went something like this: Each kid was asked to sit down at a table, where a single marshmallow was placed on a plate. The researcher told the child that if she waited to eat the marshmallow until the researcher came back, then she could have two marshmallows to eat—a real reward for a preschooler! The researcher left the room for about fifteen minutes, and—though the decision to eat the marshmallow varied from child to child—they all shared the fundamentally same experience: pure agony! Some kids stared incessantly at the treat; some turned around and couldn't even look at it; most picked it up, smelled it, took little bites out of it, and tried to get as close as they could to eating it without actually doing so. Only a third of the kids resisted the temptation and made it the whole fifteen minutes without eating the tempting indulgence. Interestingly enough, Mischel found that the children who could delay gratification in this experiment ended up having better SAT scores (about 12.5 percent higher), healthier future body mass indexes, and went on to achieve higher education; put frankly, they generally had a more promising future than those who gave in to the temptation.

While this experiment possesses a good deal of entertainment value (search for it on YouTube and find hilarious videos of kids trying to resist the sugary snack), it also teaches us how to nurture our human nature. It's true that the kids who could not delay gratification were tested twelve years later and found to be more easily frustrated, indecisive, and disorganized, while the kids who waited for two marshmallows were more confident and self-reliant. But one thing Mischel clearly articulated—and this was the purpose of the original study—is that self-control and delayed gratification can be taught. Yes, some children are more naturally prone to be unlikely to delay gratification, but they also have the capacity to learn otherwise, which invariably helps them in the future. In other words, we can teach and cultivate our behavior—nurturing positive proclivities while learning to avoid rather detrimental instincts. Thus, it would seem reasonable to conclude that the knowledge we gain about our human behavior and where certain of our tendencies come from can help us to improve as individuals, perhaps even to become holier.

In addition to this endeavor—and in light of the tension between our evolutionary instincts and our freewill—this book will address the following question: How do I live a holy life while I'm being pulled in so many directions by my biology, by my evolutionary past? And this tension leads to other questions: What does all this "science stuff" have to do with my Christian faith? In what ways can adopting an evolutionary view of human origin actually *help* my personal relationship with God and be compatible with my Christian faith?

WHERE I'M COMING FROM

Christians—and evangelicals in particular—have always had a funny relationship with science. We love it and we hate it, but more often than not, we don't quite know what to do with it. On the one hand, like most people, we love all the benefits that come from cutting-edge research—things like medicine, helpful technologies, and unprecedented communication networks. On the other hand, we often get a bit nervous when scientists tell us things that might not be found in the Bible or may, at first glance, even seem to contradict it. Take for instance the relatively new concept in human history of reproductive technologies. A few hundred years ago a man and a woman had to have sex to have a baby. In the twentieth century, people could have sex without having a baby. Now, in the twenty-first century, we have developed the technology to viably create a baby without having sex. It's understandable that a holy text that is thousands of years old did not anticipate such specific scientific developments.

Most people often find themselves in another odd relationship to science: namely, they are not scientists, though they are shaped by science every day of their lives. In the same way that most of us have no idea how our computers work or what is going on under the hood of a car, most of us don't know why human nature works the way it does. We have thoughts, emotions, proclivities, and desires. We typically take these phenomena to be given and leave it at that. While there's nothing wrong with living our lives this way, we should at least acknowledge that our lack of training in complex scientific fields probably does preclude us from commenting about their veracity. If I take my minivan to the shop for an oil change but the mechanic tells me there's something wrong under the hood, I'm at his mercy; I have to trust his assessment because, even though I am capable of driving the van and benefit from the technology every day of my life, I'm still not a mechanic and can't tell you the first thing about combustion engines. And maybe our relationship with mechanics is not too unlike our relationship with scientists—we want to trust them until they tell us something we don't want to hear, because then we have to have faith in what they're telling us, and this faith can seem costly.

My own relationship with science is not much different from what I've just described. I took scientific advances for granted and didn't think much of them, even though I have benefitted from them throughout my life. It was much the same way in my life with regard to evolution, the particular concern of this book; I grew up never really having thought about it one way or another. It's not like I was surrounded by antiscience people in my church; nor was I encouraged to cast a leery eye toward science. Yet somehow I was suspicious of evolution, perhaps owing to broader Christian culture. Years later I started asking difficult questions that led me to a place where I could accept both evolution and Christianity as being compatible truths. But accepting such doesn't mean the struggle and questions go away.

I say all this to express my affinity with the reader who may be suspicious of evolution even if cautiously accepting it. I get why it might feel threatening to one's faith. In addition to this book's purpose of demonstrating how understanding evolution can be an advantage for one's faith, I also hope the book helps some of its readers know that it's okay to ask hard questions about evolution, our faith, and how all these pieces fit together.

PICKING UP THE RIGHT BOOK

Have you ever eaten turkey bacon? It's often disappointing—especially if you're expecting to eat real bacon. Sometimes I wonder, however, if the only reason turkey bacon is so unsatisfying is because we expect it to be different than it is. It was *never* real bacon. It was always something different. We can't blame our frustrating culinary experience on turkey bacon but rather on our expectations. If you're coming to this book hoping to be convinced of evolutionary theory, then you're going to have a challenging experience. Likewise, if you're coming to this book with an entrenched six-day-creation perspective, then things will taste like turkey bacon to you. There are more appropriate books out there that are located in the middle of the evolution-creationism debate.[1] I would be equally disappointed if an individual walked away

---

[1] See Kathryn Applegate and J. B. Stump, eds., *How I Changed My Mind About Evolution: Evangelicals Reflect on Faith and Science* (Downers Grove, IL: IVP Academic, 2016);

from this book either upset because I didn't write much about the evolution debates or triumphant that whatever perspective they came with to this book—whether it was creationism or evolution—was reinforced. While I have one chapter dedicated to evolutionary theory and its compatibility with the Christian faith, I'm not totally trying to convince people of the veracity of evolutionary theory; instead, that chapter simply attempts to get the reader up to speed with theistic evolution—the notion that belief in God and evolution by natural selection are compatible—so that we can get to the heart of the book: understanding evolution can help us to be more virtuous and even holier.

While this book isn't written for everyone, it is written for a particular kind of reader: someone who doesn't get squeamish at the mention of evolution, someone who is maybe even curious about it but doesn't know why believing in that theory matters for the Christian faith. In other words, I am hoping to address an audience mostly comfortable with an evolutionary account of human origins, anthropology, and modern-day genetics who doesn't yet see its relevance to Christian life.

My intention was to write this book in a manner accessible to a broad audience of nonspecialist readers, recognizing all the while that many might not share my perspective. I hope you find this book to be welcoming, encouraging, and engaging. If you're interested in going a bit deeper, please note that I purposely relegated scholarly arguments to footnotes for those readers who may want to investigate my claims on their own. What is more, each chapter ends with a series of questions to help guide further conversation within a Christian formation group setting. I have also kept the chapters to a manageable length so it is easy for groups to read on a weekly basis.

Darrel R. Falk, *Coming to Peace with Science: Bridging the Worlds Between Faith and Biology* (Downers Grove, IL: InterVarsity Press, 2004); Karl Giberson and Francis S. Collins, *The Language of Science and Faith: Straight Answers to Genuine Questions* (Downers Grove, IL: InterVarsity Press, 2011); and Gary N. Fugle, *Laying Down Arms to Heal the Creation-Evolution Divide* (Eugene, OR: Wipf & Stock, 2015).

MY HOPES FOR THIS BOOK

In recent years there has been an explosion in the understanding of evolutionary theory and its implications for human nature. The whole concept of what it means to be human is in a state of flux. Yet in spite of all this new knowledge, most Christians either see evolution at best as irrelevant to their faith and at worst as threatening to their faith. These contrary positions often sow discord among Christian communities, precluding any sort of healthy conversation we might have. The aim of this book, then, is to bring these contrary positions into conversation with each other by having two primary goals: (1) to offer a cogent, accessible understanding of basic evolutionary concepts —from natural selection to anthropology, from genetics to the environment—while acknowledging the aspects of evolutionary theory that create tensions within basic Christian theology, and (2) to articulate what practical benefits await the Christian who adopts an integrative approach to evolution and Christianity.

From my perspective the second aim is the most important. I truly believe that having knowledge of our human ancestry can benefit our Christian life in a multitude of ways. One way in particular centers on understanding where our proclivities and desires come from. When those kids with the marshmallow were agonizing over the treat, they were acting on normal and natural instincts. As they get older, those proclivities won't go away and could possibly get even stronger. A parent that takes seriously full humanity—both our evolutionary roots, with their complicated instincts, and our sinful spiritual nature—can more effectively guide their children through life's temptations rather than denying that human history affects our behavior. We also have helpful opportunities to sit down with those in our communities who struggle with addictive behavior and encourage them to be cautious about both the physical and spiritual side of cravings and dependencies. In the same way that family medical history helps predict future medical problems, remembering our evolutionary past can provide warning signs of destructive lifestyle formations.

Besides helping us be on the lookout for negative behavioral traits, recognizing our evolutionary roots can be a positive boon for our Christian life. Our past connects us to the whole of humankind—in addition to the material and organic world—in a unique and intimate way. When we acknowledge the full picture of human origins, we can learn to nurture positive traits such as altruism, kindness, and empathy.

Over the years that I've served as both an ordained pastor in the Free Methodist Church and as a professor at a Christian university, I have met countless parishioners and students who find that accepting evolutionary theory can be empowering to their Christian faith. And I can also speak from my own experience, in which taking a serious look at human origins motivated me to get off the couch and surround myself with groups of people who could help foster holy habits in my daily Christian life.

WHERE WE'RE GOING TOGETHER

The book consists of eight chapters that are arranged in three sections: (1) "Understanding Our Biblical Lens," (2) "Understanding Our Scientific Lens," and (3) "An Integrated Approach to Evolution and the Christian Faith."

Because I realize that some readers of this book will likely be leery about the implications of evolutionary theory, I thought it would be helpful to acknowledge and discuss the difficult parts of evolution for our Christian faith early in the book. In this way I hope to put readers at ease by addressing their potential fears. While I personally hold an evolutionary creationist perspective, that doesn't mean I haven't had to wrestle with particular biblical passages.[2] Some specific concerns revolve around the historicity of Adam and Eve and the role of the fall of humanity into sin. Other concerns center on the eons of predation and death in the organic world before *Homo sapiens* ever emerged.

---

[2]Evolutionary creation was once called *theistic evolution*. Since theistic evolutionists believe that God created the universe, I think it's appropriate to be labeled a *creationist* of sorts.

One of my major priorities in this book is for the reader to understand very clearly the terminology and language that I'll be using throughout. The last thing helpful to the church is further miscommunication regarding basic evolutionary and biblical terminology. Chapter two, "Reading Scripture Faithfully," will introduce readers to both of the basic creation narratives and the common methods of interpreting them. A primary aim of this chapter is to help readers understand that everyone reads Scripture from a particular position; in other words, everyone uses a specific lens by which to understand the Bible. I consider the various creation stories (Genesis 1, Genesis 2, Job, and Isaiah) as short case studies into inductive Bible study. I take a Socratic approach in this early chapter, helping readers to ask certain questions about biblical literacy such as, Should we read all sections of Scripture the same way? If we are to read certain sections literally, poetically, or metaphorically, by what authority are such interpretations made? I hope we will come away from this chapter with a sense of the broader readings of the creation narratives within Christian communities as we look to major Christian thinkers through history who have not read Genesis literally (e.g., St. Augustine).

As previously mentioned, the first section of the book addresses what I believe to be the two main problems of evolutionary theory for the Christian faith. It is my hope that in this section I can both gain your trust and establish a context for truthful dialogue. By not shying away from what are the most difficult problems with connecting evolutionary theory to Christian theology, I work to put some readers at ease and encourage them to be open to what I have to say in the remaining chapters.

In light of this context, chapter three, "Adam and Eve, the Fall, Predation, and Death," opens with some difficult questions such as, How did Adam and Eve's sin affect the natural realm? and What do we do with the Pauline verses concerning Adam and the Second Adam? These questions, among others, are the basis of concern for many Christians when considering evolutionary theory. I draw from both historical and contemporary Christian theology in order to get at some careful answers to these questions. I will also address the other problem we encounter when integrating Christian theology with the theory of evolution: the idea that predation, suffering, and death necessarily come well before humankind and thus before the fall. It would seem that the natural world was designed to be red in tooth and claw. I first make clear that this is, in fact, a legitimate problem. It also seems that if we adopt evolutionary theory, not only do we have to assume that suffering and death came before humankind but that the death of countless species was also *necessary* for humankind's existence. I discuss the relationship between death and sin in the Bible and attempt to articulate how a Christian might begin to think about predation and death existing before humankind. I also investigate the difference between an Augustinian/Western Christian perspective on sin, death, and the fall compared to an Irenaean/Eastern Christian perspective—one that is receptive to theistic evolution.

Section two attempts to help readers better understand the scientific lens and its relationship to the kingdom, church, and the world. In chapter four, "The Nuts and Bolts of Evolution," I bring readers up to speed with contemporary evolutionary theory, discussing issues ranging from natural selection to anthropology, from genetics to the environment. I employ numerous anecdotes and stories to help describe the tapestry of life we observe daily, while at the same time investigating the free will of the natural world—where behavioral changes from generation to generation have shaped the genetics we encounter today. One major mission of this chapter is to provide a consistent vocabulary with which to dialogue in love about the theory of evolution so that community groups are able to avoid speaking past each other. In a continuation of these themes, I give a quick account of our evolutionary roots.

Chapter five, "Relating to Science," sets up the major paradigm of integration that we'll address in the remainder of the book, drawing from Ian Barbour's four approaches to the intersection of science and religion: conflict, independence, dialogue, and integration. I lay out the rich history of scientific-Christian integration—pointing to people such as Augustine and Galileo—while at the same time examining where contemporary voices (Ken Ham, Holmes Rolston, Francis Collins, et al.) would be located on Barbour's list. This chapter also encourages readers to look at science holistically, seeing the benefits of medicine and helpful technology against the dangers of scientific destruction, such as nuclear warfare or "scientism"—the claim that science itself can explain everything. By the end of the chapter I will have conveyed that evolutionary theory does not have to be antithetical to our Christian faith.

Since the crux of the book is section three, it is my hope that after we are on the same page and speaking the same language regarding the theory of evolution and the Christian faith, we can begin to understand how this integration might benefit the Christian life. This section provides practical considerations to that end. If one adopts Barbour's notion of integration, then there are several important ways that such an adoption might shape not only the way Christians interact with the Creator but also how they interact with creation.

The purpose of chapter six, "Understanding Evolutionary Theory Can Be Empowering," is to exhibit the origins of many of our natural proclivities. Here, we will be made aware of the powerful temptations that face contemporary humans. I explain how eons of natural selection helped place *Homo sapiens* in a complicated position: having caveman instincts but possessing modern technological advancements. The satirical news network *The Onion* had a funny headline that illustrates this idea: "Chemicals That Pushed Man's Ancestors to Run Down Wild Boar Flare at Sight of White Cheddar Popcorn Bag in Grocery Store." The more one is aware of this complex milieu that humans find themselves in, the more empowered they can be to overcome tempting biological and environmental urges.

Chapter seven, "Having Evolutionary Roots Isn't Just Baggage: Pursuing a Holistic Understanding of Redemption," encourages readers to embrace an integrative approach, which leads to a more holistic understanding of redemption that engenders a flourishing Christian life. In fact, I wish to make clear in this chapter that there are numerous *positive* evolutionary characteristics that humans have and can contribute to a movement toward altruistic practices and a holier way of life. Adopting evolutionary theory does not necessarily impede our movement toward a deeper life of faith and commitment to the kingdom—it may, in fact, embolden us. Redemption informed by evolutionary theory becomes, then, not only the story of the redemption of one's soul but also the redemption of one's mind and behavior—a redemption that invigorates the fine-tuning of one's virtues.

Chapter eight, "Nurturing Natural Virtues Toward Healthy Community," takes a precise look at how church formation groups can cultivate virtuous and holy living. This chapter is practical in nature and provides historical and contemporary examples of groups who—with the knowledge of natural and spiritual behavioral influences—developed healthy habits that resulted in virtuous and holy living. I hope to encourage readers to pursue further conversation in whatever communities they reside.

It's my hope that this book can help readers lead a more fulfilling and virtuous Christian life. When you are finished with the book, it is my wish that you might be better able to articulate where dangerous instincts come from—and how to overcome such proclivities—as well as to encourage the drives that positively nurture holy living. Temptations over eating marshmallows aren't going away anytime soon. Hopefully, we can learn about where we come from in order to positively shape our Christian lives.

PART ONE

UNDERSTANDING OUR BIBLICAL LENS

2

READING SCRIPTURE FAITHFULLY

WHEN I WAS A PASTOR, I would get questions about difficult Bible passages. And maybe it was my lack of confidence or maybe it was my seminary training (or probably a little of both), but I would always defer to church tradition for the answer—either my own Free Methodist tradition or the early church fathers and councils. Who am I to disagree with *them*? Currently, I teach theology and philosophy at a Christian university, and I am surrounded by intelligent and godly Bible professors. So now, when students come to me with complicated Bible questions, I get to defer to my colleagues, which is quite a relief. I've found that my colleagues answer our students in much the same way as I counseled my parishioners—in deference to our faith, pointing them to what church tradition says about the passage while using their reason and their personal experience with God as their guide.

Reading the Bible is complicated. It's not often straightforward, and many people suggest various methods of interpretation. The purpose of this chapter is not to propose some new method for reading Scripture. Nor in it do I propose to elucidate all the intricate biblical passages that have been troubling scholars and clergy for generations. Instead, what I hope to express is how, by leaning on the practices established in the church over the last two thousand years, we can come to a robust and historical way of reading Scripture.

When approaching any book that purports to speak on Christian ideas, the critical and careful reader ought to pay attention to how the author views Scripture. And so, to be forthright with my readers, I've placed this chapter immediately following the introductory chapter. It's important to me for readers to know that I'm orthodox in my theology, steadfast about the authority of Scripture as interpreted through Christian tradition, and an advocate for the place of reason and personal experience with God when reading the Bible. In this way I'm both Wesleyan and little *o* orthodox. Thus, I believe this book is for the majority of practicing Christians, from mainline Protestants to evangelicals, from Roman Catholics to Eastern Orthodox.

SEEING THROUGH A LENS DARKLY

All humans see the world upside down. We view an image in the world, the light passes through the lens in our eyes, and then the image is flipped on its head. The brain takes those upside-down images from each eye and does something marvelous: it combines the images, flips them right-side up, and makes sense of the world. It's an amazing process that exhibits how remarkable the human body is.

George Stratton was one of the world's first experimental psychologists conducting experiments in the early twentieth century. He was so fascinated by what was happening within the eye that he created an elaborate experiment. So, if you're keeping track, the image enters our eyes, gets flipped by them, and then gets flipped another time by our brain. Stratton, however, wanted to flip that image *again* using a set of special goggles he developed in order to test whether or not the brain could adapt and flip the image right-side up one more time. If the experiment worked, he would have proof that we get comfortable seeing the world a certain way after a while.

Stratton wore these goggles for a full week. At first, he was nauseous and disoriented. He fell repeatedly, and it was as if his body wasn't his own. The first few days were full of falls and sickness. But after the fourth day, he began to adapt, and after only seven days of wearing the upside-down glasses, Stratton was able to see the world right-side up again—his eyes had adjusted to a new normal. Through this experiment Stratton proved that over time he was less aware of his goggles and how they shaped the way he saw the world.

In a philosophical sense Stratton's experiment applies to the ways we variously experience the world; we are all looking at the world through lenses—and we're so comfortable with them that we often don't know we're wearing them. We're conditioned by our surroundings and habits to think and see certain ways. The same can be said for how we see and read Scripture—nobody reads Scripture without wearing certain lenses; it's impossible not to do so. But that's not necessarily a bad thing, especially since it's a basic reality of our creatureliness. We must, however, acknowledge that there are lenses that distort the truth of Scripture and those that clarify it. In many ways, doing theology is the practice of working out in community which lenses we ought to put on and which we ought to take off.

Everyone is shaped to read Scripture in particular ways whether we realize it or not; in much the same way that worldviews are formed, our approach to reading Scripture is often initially formed in a pretheoretical fashion—that is, it is formed before our awareness of such. Just as Stratton's brain adjusted to the effects of his special goggles, so we often come to look at Scripture through lenses we're unaware have been placed on us. These lenses we don as we read Scripture are like a prism that splinters a ray of light; God's truth is the pure beam of light, and we desire to work our way back toward that pure light. Yet we see that light through a glass darkly, and so it comes to us splintered in greens, blues, reds, and yellows—beautiful though imperfect. Because we do not yet see that light clearly, we must work carefully and diligently to avoid heresy and heterodoxy. The way almost all orthodox Christians accomplish such interpretive and theological work is through an approach that appeals to Scripture, tradition, reason, and personal experience with God. These four points are like the splintered light—they are all necessary parts of the pure beam, and though we come to them separately, they work in union. Indeed, various Christian traditions differ on the order of importance placed on each point; in fact, some say there are only three points—some say separating the points is unnecessary. But for the purposes of this discussion, we're appealing equally to the four points, a common practice among orthodox Christians.[1] And while there is a place for debates about the order and importance of each point—for example, my own Wesleyan tradition places the greatest emphasis on Scripture—this book is not it. What's important for our discussion is that in one way or another all Christians see and interpret their faith through these four points.

And so, when we read and interpret Scripture, we do well to acknowledge that we do so from a particular perspective. Whether we realize it our not, we read Scripture out of church tradition, reason, and our personal experiences. It's nearly impossible, and probably dangerous, to read Scripture in isolation from these. For example, when we read the words "God the Father," we aren't expected to ignore the connotations that God is like our own earthly father. But very quickly we understand how reading this appellation out of only our own personal experiences is inadequate. Some of us have had great earthly fathers; and for us, when we read "God the Father," we understand God to be endearing, full of love, supportive, selfless, and so on. For others, coming across proclamations of God as *Father* in Scripture might conjure childhood trauma, past neglect or abuses, or simply distance and abandonment. Whether we're aware of this or not, we read these sacred words of Scripture through the lens of our own experiences.

So what do we do if Scripture is talking about the Father's love and we had an earthly father who was anything but loving? We go to the other points of light to help clarify our vision: church tradition is

---

[1] These four lenses are what the Wesleyan tradition calls the "Quadrilateral." While the Quadrilateral is distinctly Wesleyan (in name only), it has its roots in Anglicanism, Catholicism, Eastern Orthodoxy, and most other Protestant traditions.

clear—all through the centuries—that God's love is pure, unadulterated, faithful, and endless. Our ability to reason, another point of light, can help us rationally process how tradition and Scripture clarify that our experience of our earthly father is distinct from our heavenly Father, who Scripture expresses as pure love. Many of us go through this process instantaneously and thus don't necessarily realize that we're coming to an interpretation of Scripture via tradition, reason, and experience. Others methodically focus on these points to peel back the lenses we use to read Scripture. Thus, in this chapter I hope to spend some time elucidating the lenses that shape my interpretations.

I recognize that this might be a new concept for many readers, but consider the alternative to reading Scripture in this way. When we claim to read in isolation—that is, only using Scripture to interpret Scripture—we are either (1) reading through a tradition, reason, or experience that we do not wish to acknowledge, or worse, (2) we, in isolation, are being the final authority of our own scriptural interpretations. The latter is the road that the Mormons and Jehovah's Witnesses have traveled down. A third option, however, is more subtly dangerous: our local church or pastor is the sole interpreter of Scripture. One of the reasons this multilensed reading of Scripture is so vital to the preservation of orthodox Christianity is that it rests on a foundation of countless church leaders throughout the centuries and not merely on the interpretive faculties of one person or congregation.

I write this from a position of genuine love for Scripture. And it is out of this love that we ought to protect the Bible from heretical or heterodox beliefs that most of Paul's writings in the New Testament warn us against. But one might rightly question whether using tradition, reason, and experience as a lens to understand Scripture could *itself* lead us to bad theology. Take the resurrection of Christ or the future resurrection of the saints, for example. At first glance both our reason and experience of the world would tell us that dead things don't rise from the dead—and we would be correct in trusting these faculties. But taken together with Scripture and church tradition, which tell us otherwise, we must not dismiss the miraculous. Looking more closely at our reason would also tell us that we haven't learned everything about the world, and we will likely never learn everything there is to know. Additionally, a deeper dive into our experiences shows us that we haven't experienced all things, nor will we ever. But we have whole councils set aside to debate and ultimately affirm the resurrection of Christ and of us. Concepts such as the Trinity, the virgin birth, and the new heaven and new earth fall under the same parameters. We defer to tradition and Scripture, even though the resurrection or the Trinity doesn't seem to make "rational sense," in the traditional use of the phrase.

HAVING A HEALTHY LENS

Some lenses provide a healthier perspective than others. I remember when I took a summer job in Kentucky putting flat rubber roofs on office buildings. My first day on the job I got up on a roof, in the hot southern sun, and my boss, Daniel Boone Logan—no joke!—said, "Didn't you bring tinted safety goggles?" I said, "No, but I do have these regular safety goggles." "O-o-oh buddy," Boone said in a prolonged Kentucky drawl, "it's gonna be a long day for you, son." Man, was he right. After an excruciating day of seeing the sun reflect off of that bright white roof, I went home and my eyes felt like they were going to fall out of my face. Yes, I was wearing safety goggles, but I was wearing the wrong *kind* of goggles with the wrong *kind* of lenses. The next day, you better believe I brought some tinted glasses.

There are healthy and unhealthy lenses through which to read Scripture. I was talking to a Bible scholar at my university about this very issue, and he made a convincing case for reading Scripture with what he called an *incarnational* lens, one that reads Scripture through the interpretation of church tradition, reason, and experience and not a *dualist* lens—by which he meant to mean nonincarnational and dualistic—and thus views Scripture as inactive and distant. This revolutionized my thinking. Think about it: most of the New Testament—pretty much everything that Paul wrote—is a critique of dualist ideas and an affirmation of the incarnation of Jesus Christ. Dualism became popular around the time of Christ and has its roots in Platonic philosophy. In a nutshell, it claims that the body is bad and the spirit is good. Paul didn't like this because it undercut the incarnation and resurrection of Jesus (among other fundamental doctrines like the virgin birth and the new earth). Contrary to Scripture's vision, dualism declared the spiritual to be perfect and the physical imperfect, messy. What is one to do, though, with the central place of the incarnation in the gospel: God came down into this messy world and dwelled among us. What a beautiful picture of love and grace.

We want Scripture to be cut and dry. We don't want any ambiguity in Scripture because we believe that Scripture is authoritative. And while I believe Scripture to be authoritative, I also believe we have to be careful not to let authority become idolatry. Scripture is living and breathing and incarnational. It was written through people (body) by divine inspiration (Spirit). But it would be dangerous to confuse authority with idolatry by making the Bible itself into some kind of god—such worship of Scripture is called biblicism. And we must remember that Scripture is a *revelation*, pointing us toward Christ, who is truth. The Bible isn't the way of salvation; it is a message that points to the way of salvation. It's also not powerless; instead, it is a dynamic presence in both the past when it was written and our present reality.

The incarnational reading of Scripture my colleague described is messy and takes work to search church tradition, reason, and experience. But in the long run, an incarnational reading of Scripture is healthier and truer to the authors' original intent and to our contemporary experience. It's messy only because people are messy. Take the historical church councils as an example. The early church fathers would read Scripture, have differing opinions, and then work out these differences at a council to determine what the proper orthodox belief was. This process might sound scary and was certainly messy, but it's also a beautiful representation of how God works *through* his people. It doesn't get more incarnational than that.

All of our main Christian doctrines were worked out by this process: reading Scripture, studying tradition, checking our personal experience with God, and reasoning it out in an ecumenical council. From the Trinity to the dual nature of Christ, the Christian faith is used to this process of establishing doctrines. In fact, that's one of the things that separates the Christian religion from Islam. In Islam, the belief is that the Qur'an was written in heaven and is an eternal document. So a command in the Qur'an is an eternal command.[2] In Christianity the Bible is viewed as a historical document—one that is *inspired* by God but not *written* by God. This means that we interpret commands in their historical context. For instance, when we read 1 Samuel 15:3—"Now go and attack Amalek, and utterly destroy all that they have; do not spare them, but kill both man and woman, child and infant, ox and sheep, camel and donkey"—we don't read that as a command for contemporary Christians to continue pursuing the Amalekites today. Instead, we read that passage as a specific command to a specific people in a specific time.

Wouldn't our faith be easy if such passages were universally applicable, making truth unmediated, direct from God, rather than being revealed to us through the work of interpretive processes, through the medium of people with all of our foibles? But God asks us to be interpreters, to be flesh and blood, to be fully human. We need only look to the essentially human nature of the gospel for proof that humanness is central to God's plan for our lives in this world—the incarnation, the virgin Mary, the bodily resurrection of Christ, human authors of

---

[2]To that point, because Islamic scholars believe it is an eternal document, it is therefore unable to be translated, as it is believed to be whole—in its entirety—a book written in Arabic without beginning or end.

Scripture. The same Bible professor was telling me how he brought students to the Island of Patmos, where John wrote his letters. When they entered the cave that has now become altered to make a small chapel, the students were very disappointed. "Why are you all dissatisfied?" He asked. "Because we were hoping it would be just a cave, not a church." And there's the crux of the problem. We want Scripture to be just the text, unmediated through any other lenses. But that's not how the incarnation worked, nor is it how an incarnational reading of Scripture works. A healthy lens by which to understand Scripture is viewing the holy text through the prism of church tradition, reason, and experience in order to fully unpack the truthful message that it holds.

HAVING A HEALTHY CONTEXT

Sometimes I get asked whether or not I take the Bible literally. Usually, this question arises because the askers are about to claim that they do and are doubtful that I do. But what an odd question, given that I've never met a person who takes the entire Bible literally. Not a single person. Jesus says in Mark 9:47 that if your eye causes you to sin through lust, then you should pluck it out. I was walking with my father-in-law in Egypt through some catacombs of the early church, and inside one of the caves was a mural of St. Simon the Tanner painted on the wall. In every portrait of Simon the artist portrayed him as having only one eye. My father-in-law said that it was because Simon took the Mark 9:47 verse literally. We both paused and thought about it for a moment. Then my father-in-law, who himself was a missionary in the Middle East for forty years, turned to me and said with a wry smile, "Well, I guess he only ever lusted once, huh?" There are only two options here—either Jesus was speaking metaphorically or we all fail to act on one of Christ's clearest commands. But since churches aren't hosting eye-gouging ceremonies, I can assume we take Christ's words to be metaphorical. And I assume this because tradition and reason are at work in my life.

And what about Matthew 13:3? Jesus taught a group of people through parable, beginning "A sower went out to sow." Did this event with the sower historically and literally take place? No, of course not. But Jesus was telling a parabolic story to teach the crowd about salvation. What's more, when Jesus calls the Pharisees a "brood of vipers" in Matthew 3:7, he didn't literally mean that they were snakes. Jesus was using hyperbolic language to make a point. Why would we assume that the authors of the Bible were only able to use literal language when giving an account? They were completely capable of moving between allegorical, historical/literal, spiritual, and moral language, just as we are today.[3] Interpretation is necessary when we read Scripture, and so claiming that a literal reading is the only true way of reading Scripture is at best specious and at worst un-Christian.

So when people ask, "Do you take the Bible literally?" they are really asking a cultural question, not a religious question. In other words, they're asking what my mode of interpretation is when I read the Bible. As mentioned before, we all read and interpret Scripture through certain lenses. Almost all of us preference a certain way of reading Scripture over others—but we must strive to allow our reading and interpretation of Scripture to be shaped in a healthy way by our church tradition, reason, and experience.

Augustine of Hippo was passionate about this subject. This saint, whose thoughts have had a lasting influence on theology and philosophy, acknowledged that some passages of Scripture are difficult to interpret and, in fact, often lead to various interpretations among the most careful students of Scripture. He says,

> In matters that are so obscure and far beyond our vision, we find in Holy Scripture passages which can be interpreted in very

---

[3] We should note the early ideas of this come from Irenaeus to Nyssa as well as De Lubac's *Medieval Exegesis* and the fourfold reading of Scripture. See Gregory of Nyssa, *The Life of Moses*, Classics of Western Spirituality (New York: Paulist Press, 1978); and Henri de Lubac, *Medieval Exegesis: The Four Senses of Scripture*, Ressourcement (Grand Rapids: Eerdmans, 1998).

different ways without prejudice to the faith we have received. In such cases, we should not rush in headlong and so firmly take our stand on one side that, if further progress in the search for truth justly undermines this position, we too fall with it. We should not battle for our own interpretation but for teaching of Holy Scripture. We should not wish to conform the meaning of Holy Scripture to our interpretation, but our interpretation to the meaning of Holy Scripture.[4]

And so we can see from the earliest days of the established church, we have wrestled with how to interpret Scripture. And since these early days theologians have been careful to insure that their interpretations conform to Scripture, not the other way around.

If you're familiar with the renowned atheist and critic of religion Richard Dawkins, you know that one of his naive critiques of Christianity and Scripture is actually a straw man attack and goes something like this: To the extent that you can find the good bits in religious scriptures, you have to cherry pick. You search your way through the Bible and you find the occasional verse that is an acceptable profession of morality and you say, "Look at that. That's religion," and you leave out all the horrible bits. Of course, Dawkins often acts in a reductionistic manner, and he doesn't typically acknowledge the nuance of biblical interpretation—that to read Scripture one must do so through the lenses of tradition, reason, and experience. Ironically, Dawkins's shallow critiques are not unlike that of the Christian who reads Scripture rigidly or claims to do so only literally. We might respond to Dawkins with words we ourselves should heed—if he took certain passages in historical context or reasoned through church tradition, then he might come to a fuller interpretation of those passages—a message that is good for us as well.

Certainly, we don't have to make historical biblical criticism the bad guy. In fact, it can help protect the faith. For the purposes of this book

---

[4] Augustine of Hippo, quoted in Alister E. McGrath, *Science and Religion: An Introduction*, 2nd ed. (Malden, MA: Wiley-Blackwell, 2010), 5-6.

I will strive to interpret Scripture in humility, looking to God's natural revelation, Christian tradition, and my own experience to guide the words that follow. I cannot write this book without acknowledging that we place a certain amount of faith in science, which often reveals how the world works. Science itself, however, cannot stand alone as an arbiter of truth. Thus science must also be understood through the proper interpretive framework.

It is my hope in this book to be forthright and careful, to acknowledge the complex yet simple nature of Scripture. Of course, we shouldn't balk at the fact that interpreting Scripture is complicated—most things in life are more complicated than they often appear. Life itself is complicated. Why should we expect reading Scripture to be uncomplicated?

Because I wish to display this chapter's perspective as normative throughout Christian history, I'd like to end with an extended quote by Augustine of Hippo (AD 354–430) from his letter titled "On the Literal Meaning of Genesis."

> Usually, even a non-Christian knows something about the earth, the heavens, and the other elements of this world, about the motion and orbit of the stars and even their size and relative positions, about the predictable eclipses of the sun and moon, the cycles of the years and the seasons, about the kinds of animals, shrubs, stones, and so forth, and this knowledge he holds to as being certain from reason and experience. Now, it is a disgraceful and dangerous thing for an infidel to hear a Christian, presumably giving the meaning of Holy Scripture, talking nonsense on these topics; and we should take all means to prevent such an embarrassing situation, in which people show vast ignorance in a Christian and laugh it to scorn. The shame is not so much that an ignorant individual is derided, but that people outside the household of faith think our sacred writers held such opinions, and, to the great loss of those for whose salvation we toil, the writers of our Scripture are criticized and rejected as

unlearned men. If they find a Christian mistaken in a field which they themselves know well and hear him maintaining his foolish opinions about our books, how are they going to believe those books in matters concerning the resurrection of the dead, the hope of eternal life, and the kingdom of heaven, when they think their pages are full of falsehoods and on facts which they themselves have learnt from experience and the light of reason? Reckless and incompetent expounders of Holy Scripture bring untold trouble and sorrow on their wiser brethren when they are caught in one of their mischievous false opinions and are taken to task by those who are not bound by the authority of our sacred books. For then, to defend their utterly foolish and obviously untrue statements, they will try to call upon Holy Scripture for proof and even recite from memory many passages which they think support their position, although they understand neither what they say nor the things about which they make assertion.[5]

In other words, Augustine is clearly stating that for the sake of evangelism and reaching the lost we need carefully to read Scripture through church tradition, reason, and experience. Augustine is telling us that disregarding this advice is not simply embarrassing for the church but actually hinders nonbelievers from taking us seriously—ultimately turning them away from the faith.

WHAT IS GENESIS ABOUT ANYWAY?

Our lives were in a great place. We were finishing our educations, ready to make a big move to a new state and start the lives that we had been trying to build for the last seven years. My seminary training and my wife's graduate work were ending and we had lofty plans: paying back student debt and eating more than beans and rice. So my face went a bit flush when I heard her say the heavy words, "Matt, I think I'm pregnant."

---

[5]Augustine, *The Literal Meaning of Genesis*, trans. and ed. John Hammond Taylor (Mahwah, NJ: Paulist Press, 1982), 43.

"No, you're not," I confidently assured her. "That's impossible."

"Um, yeah I think I am."

So we drove at ten at night to the nearest Walmart to buy a pregnancy test. That was one of the scariest drives of my life. It's not that we didn't want kids someday, but it wasn't good timing for us right then.[6]

When we got home we immediately went to the bathroom. My wife took the pregnancy test, and we waited, eyes fixed on the stick for what seemed like an eternity. Never have the symbols of addition and subtraction mattered so much to me as they did at this moment. Slowly, a plus sign emerged and my wife said, "Matt, I'm pregnant!"

"No, you're not," I confidently assured her, again. "That's impossible."

"A plus means that I'm pregnant, and there's a large plus sign on the stick."

"No, a plus sign means that you're positively *not* pregnant." I said this not as some kind of terrible joke but rather out of my situational self-delusion. My wife, always the realist, calmly told me that I was wrong while I fumbled at the box trying to decode the mysterious hieroglyphs.

"It's like I need the Enigma machine to crack these instructions."

"Matt, you're being ridiculous."

"Our pregnancy test came with two sticks, right? So give me the other one and I'll take it—since we know I'm not pregnant—and then we can compare the two," I said. I was so convinced that she simply couldn't be pregnant that I had to find an alternative reading to the positive pregnancy test. As my long-suffering wife watched me pee on the pregnancy stick, she had to have wondered what kind of father I would make.

Gazing alone at the pair of pregnancy sticks resting side by side on the counter, and seeing my test slowly emerge as a solid minus sign, I looked up at my patient wife and said, "Oh no, I think you're right, you're *so* pregnant!"

---

[6] A quick aside. This experience gave me a lot more empathy for single women with unwanted pregnancies and their difficult situations. I could only ever know a fraction of their fear, and it terrified me. This kind of empathy has been a welcomed addition to my pro-life position.

Sometimes I think that we Christians have been so convinced of a certain reading of Scripture that we are unable to accept the evidence before us. We can so forcefully convince ourselves of something that the psychological denial of even a pregnancy test can trump reality. Unfortunately, there's no pregnancy test proclaiming that natural selection is the method by which God formed the universe, but there is an overwhelming amount of evidence that points in that direction. And for those of us who are troubled by contemporary readings of the Genesis narrative—given the multitude of disciplines that point toward evolution, from physics to chemistry and from astronomy to biology, I believe there to be an honest and authentically Christian way through this conundrum.

One of my favorite professors at seminary was Robert Mulholland, who wrote a fantastic book, *Shaped by the Word*, in which he discusses the difference between reading the Bible for information and reading it for formation—specifically spiritual formation.[7] Mulholland was not at all saying that the Bible has no information to give—quite the contrary. We can find all sorts of information about who God the Father is, our doctrines of the person of Jesus Christ, and the workings of the Holy Spirit. But that information doesn't give us any indication about how to have a relationship with this God. On the contrary, the Bible only helps us deepen our relationship with God when we read it for formation. Instead of reading the Bible to find out the specific dates of scientific events—which is anachronistic given that modern science didn't begin to take shape until the sixteenth century—we should be reading the Bible how the original writers intended: as a series of manuscripts to help a persecuted group of people worship the one, true God.

Any reading of Genesis will invariably have to deal with the conflict between the two creation stories found in both Genesis 1 and 2. The

---

[7]M. Robert Mulholland, *Shaped by the Word: The Power of Scripture in Spiritual Formation*, rev. ed. (Nashville: Upper Room Books, 2000).

former expresses the idea that God created the world in six distinct days, the latter indicates that God created the world in one day. To put it frankly, it doesn't seem as if the author of Genesis intended us to read the text for information but rather for formation. We simply can't rectify the two different creation accounts when reading for information. Yet a spiritually formative reading of the text—where we try to open up space in our Christian lives to be changed and challenged by the text—can offer a truthful and faithful account of God's compassion and love.

When I read Genesis I see a God who called creation inherently good. This assertion is fundamentally different from how Islam sees creation: as fundamentally suspect and possibly even flawed. I imagine an author trying to express the fact that humankind is special. I see that God himself walks with creation, talks with the humans, and endows humankind with his image, that humankind has a responsibility to creation and to God and hasn't quite lived up to that responsibility. I understand the author to be expressing the idea that humankind wanted the benefits of communing with God without the obligations of communing with God, that they stumbled time and again—first Eve, then Adam, then their children, then Noah, even the patriarchs of the faith.

If I'm primarily reading the Bible for formation rather than information, I am reminded to keep first things first: allowing my reading of the text to represent as fully as possible the author's historical and literary context. I am aware that this interpretation will ruffle some feathers. Yet I hope by this point in the book it is utterly clear that I hold both Scripture to be authoritative and nature to be God's creation. As Christians we must believe that Scripture and nature are compatible—that their narratives are interwoven.

Therefore, if the Genesis narrative of Adam and Eve isn't an explication of evolution (which it clearly isn't), then what is it? It's about separation from God, it's about idolatry, and it's about understanding our relationship with God as Father and us as children. These notions permeate the Genesis narrative; any other reading, in my opinion, sidetracks us from the author's context. In a very real way the Bible is the church's book, not the historian or scientist's. The early church has long since dealt with this and saw the primary purpose of Scripture to be forming our spiritual lives. In a very real way, when we look to the Bible as a text whose primary purpose isn't spiritual formation, then we are placing on Scripture requirements that are unreasonable and misinformed. For instance, in the New Testament John moves narratives around—away from chronological order—performing redaction (i.e., editing) for theological reasons. When comparing John's Gospel to the Synoptic Gospels in Matthew, Mark, and Luke, it's clear that John's agenda expresses correct theology, not chronology. He's concerned with formation, not information. Luke's genealogy provides another example of such. Since he is a Gentile, it's important to him that Jesus' genealogy goes back to Adam, the proto-parent of all people, not just Jews. He redacted the story in order to show that Gentiles were part of God's plan too.

Ultimately, if we are comfortable doing what the writers of the Bible were doing, and we are comfortable doing what the early church fathers were doing, then we should be comfortable reading Genesis with both spiritually and theologically formative lenses rather than chronological or literal ones. Yes, humankind has sinned and has a severed relationship with its Creator. How that came about and how that fits with the billions of years of organic life on earth is not what early Genesis is interested in. The author is advocating that the readers mend, heal, and perfect their relationship with the Creator.

3

ADAM AND EVE, THE FALL, PREDATION, AND DEATH

TODAY AS I SAT DOWN TO WRITE about suffering, I opened up the news to find that there was a shooting in Las Vegas—the worst mass shooting in US history to date, with hundreds wounded and scores dead. I debated whether or not to click on one of the many videos that were appearing online. *Why?* I thought, *I've seen violence. I've experienced violence. I'm inundated with this destruction in the news every day. Why should I click on another video?* But, as I sat there, I thought about the fact that each of these horrific incidents is fresh and new to the people they involve—despite how much violence has taken place in the past. When one sees the footage from this event, the rattling of gunfire from one maniacal US citizen to innocent others sounds in the thousands. Bodies tumbling, people running, scores bleeding. So often the suffering seems distant as we read news headlines each morning about someone else's starving kid in someone else's country. Yet, despite the abstract suffering of people I don't know in places I've never been to, I'm moved. Thus, as I set out to write about suffering—even in the midst of this terrible, horrifying news—I've got to come to terms with the reality that such suffering has always been around us, since the beginning of time.

By admitting that predation, the act of one animal preying on another, and suffering have been with creation since its inception, I risk losing some of my readers who may already doubt the validity of evolutionary theory, which I believe to be true. In an effort to demonstrate to the reader that I desire authentically to move toward truth—especially a truth that can incorporate natural revelation—I need to air the dirty laundry of evolutionary theory. I believe that evolution by way of natural selection is, in fact, compatible with the Christian faith. However, that belief does not mean that there aren't any problems with *how* evolutionary theory fits with our Christian perspective. But just because something is complicated doesn't mean it's not valid.

By addressing what I believe to be some of the main problems of evolutionary theory for the Christian faith, it is my hope that we can establish a context for truthful dialogue.[1] Such a truthful dialogue requires that I not shy away from the most difficult problems that arise from connecting evolutionary theory to Christian theology. At the same time I will work to put readers at ease and encourage them to be open to what I have to say.

LIFE IN THE GARDEN

The lessons most often taught to children in Sunday school classes come from the Old Testament. I didn't fully realize this phenomenon until I had children of my own. When coming home from church, I would ask them what they'd learned in Sunday school, and they would invariably say something to the effect of, "Jonah, Dad. We always talk about Jonah." I get it. Narratives are a lot easier to teach children than,

---

[1] Sociologist Elaine Howard Ecklund argues that some of the main problems with evolution for religious people are understanding human uniqueness and the manner in which God is involved with creation. I address the former throughout this book. An answer toward the latter can be found implicitly in my insistence that God is Creator and incarnate in nature itself through the person of Jesus Christ. This is no mere nod to the problem of divine action—or how God interacts with the world—but rather recognition that a divine being can't be more active in creation than becoming part of it. See Elaine Howard Ecklund and Christopher P. Scheitle, *Religion vs. Science: What Religious People Really Think* (New York: Oxford University Press, 2018).

say, the meaning of *wisdom* in the book of Proverbs. This teachability of narratives, while not wholly responsible, contributes to the firm belief in young people that these Old Testament narratives are all literal. As we discussed previously, some passages are surely literal and some are surely figurative, and we must look to theological tradition to help us distinguish the difference.

The narrative of Adam and Eve exemplifies the complexity of this dance between literal and figurative readings of Scripture. It is layered with imagery from ancient Near Eastern culture and informed by pre-biblical texts like the *Enuma Elish*. Undoubtedly, both in early Christianity and more notably during the nineteenth and twentieth centuries, Adam and Eve have been a source of controversy. Their scriptural narrative raises substantial questions: Are Adam and Eve real, historical figures? Is this particular Genesis narrative literal or figurative or a sort of amalgamation of the two? If they aren't historical figures, what do we do with the theological concept of the new Adam found in Romans? If they are historical figures, what do we do with the near unanimous consensus among biologists, geneticists, and anthropologists who suggest otherwise?

Indeed, Adam and Eve are a problem for belief in evolution. But they've also been a problem for theologians and Christians throughout our faith's history. Take many of the early church fathers, for example. From Origen in the second century to the Cappadocian Fathers in the fourth, there have been various interpretations of who Adam and Eve were and how they related to God. Those who helped the early church form the doctrine of the Trinity—Basil the Great, Gregory of Nyssa, and Gregory of Nazianzus—all had varying interpretations of not only Adam and Eve but also the entire Genesis creation story. Second-century theologian Tertullian, for instance, argued that *every* human who is formed in the womb, not just the first couple, is specially created and connected to God. In a sense Tertullian argues that Adam and Eve's creation and the *imago Dei* in them isn't particularly unique;

rather, he claims that *all* humans are endowed with such grace and are specially made. If the early church fathers were all over the map on Adam and Eve—see Irenaeus and Augustine for quick examples of this—then it seems acceptable that we can be open to a more generous interpretation of Genesis.

Consider with me what many, but not all, biologists suggest might be true concerning the origins of humankind: there was, as most contemporary biologists agree, a single male human from which all living humans are descended.[2] Also imagine with me that biologists are correct, through the same DNA evidence that we readily stand by in the criminal justice system, that there was a single human woman—whom they call Mitochondrial Eve—from whom all *Homo sapiens* hail.[3] So far, so good. While all humans alive on earth hail from two individuals, they were not your commonly thought of garden couple. But, giving these biologists the benefit of doubt, and for the sake of

---

[2] It is extremely important to emphasize that while scientists are in agreement, by and large, that humankind developed through a process of natural selection, there is no real consensus about how *exactly* that happened. Several theories include no single human origin; some theories include one man and one woman who lived at the same time (see John H. Walton, *The Lost World of Adam and Eve: Genesis 2-3 and the Human Origins Debate* [Downers Grove, IL: InterVarsity Press, 2015]). Although it should be noted that these theories that include one Adam and one Eve are typically from Christians seeking to integrate biology with the Bible—using the Bible as the lone source of evidence for a physical Adam and Eve. The theory I have stated above comes from Darrel Falk (see Darrel R. Falk, "Human Origins: The Scientific Story," in *Evolution and the Fall*, ed. William T. Cavanaugh and James K. A. Smith [Grand Rapids: Eerdmans, 2017]).

[3] It is important to note that many biologists and anthropologists believe these two figures to be separated by several thousand years and several thousand miles, and surrounded by numerous other humans. While some recent studies suggest that there could have been some overlap, they may have not been, according to these scholars, in the same place at the same time. For more on this, see Falk, "Human Origins." Also displaying the idea that there is no certain agreement about the origin of humankind, it is entirely possible that there were more recent common ancestors of everyone alive today. This nuance is the distinction between genealogical and genetic ancestry that has become important in some studies (see S. Joshua Swamidass, "The Overlooked Science of Genealogical Ancestry," *Perspectives on Science and Christian Faith* 70, no. 1 [2018]). Thus, science is not necessarily ruling out a historical Adam and Eve. All this to say, that while many scientists agree on many ideas within evolutionary theory, the origins of humankind are still contested. For our purposes, I assume *that* God created and take an indifferent approach as to *how* God created.

following the argument, also imagine with me that what the consensus of Christian theologians profess to be true is also correct: that God—ever being active in his creation—created humankind to be unique from other animals, fundamentally good and complex, and the pinnacle of his creation.

If, for the sake of conversation, we affirm not only the heart of what Christian theology has professed over the ages (that humans are *imago Dei* creatures) while acknowledging the differing opinions between theologians but also what contemporary science is telling us, is there any way the historical Adam and Eve don't simply blow away like so much dust in the wind? I realize that even doing so for the sake of conversation may make readers uncomfortable—so much seems to be at stake even in the consideration of the idea. Most of us have been raised with a theology that posits Adam and Eve need to have been a real, isolated couple not separated by years and distance as some scientists suspect.

But, for a brief moment, consider with me this possibility. For our Christian faith, what if it didn't really matter about all the details concerning *how* God created but rather *that* God created—regardless of his methods. It is not theologically necessary that the early Genesis stories of the creation of Adam and Eve are literal and historical. It is, however, theologically necessary that God was active in creating, an idea that I stand with the Christian church in believing. So consider what it might be if the narratives in Genesis provide for such a generous reading? In fact, the Genesis account of the creation of humans must be interpreted in the light of the New Testament, which means we come to see God's bestowing of the *imago Dei* to be, ultimately, tied to the fully human and fully divine person of Christ, who John tells us is the Creator. By considering the creation narrative in light of the person of Christ, it might not be that the bestowing of the *imago Dei* needs to happen to a single couple at a single moment and can't, in fact, be placed with *Homo sapiens* across history. In fact, we believe very strongly in the Christian faith that God has been and will always be working with every human to draw each of us closer to the Trinity. So it shouldn't scare us that the *imago Dei* moment might be a multifaceted event.

In fact, literal interpretations of the events in early Genesis really weren't prominent until the eighteenth century when the quest for historical Jesus movement began. Before then, there were copious interpretations of the historicity of Adam and Eve. But in the eighteenth century a group of academicians attempted to throw away anything in the Bible that was not "natural." In a way they were trying to argue for a deist God and not a triune God that included the person of Jesus, ripping out sections of the Bible (which some quite literally did). In response the Christian church harshly—and rightly—rejected such an interpretation. Yet, in haste, many of the laity took a defensive posture and began requiring particularly literal interpretations as litmus tests against a perceived anti-Christian cultural threat. Such interpretations, however, do not align with those of the early church fathers. Thus, in the reactionary work of certain Christians at a certain place in time, the *purpose* of the narratives are lost in order to defend against an attack on Christ's divinity. And a false dichotomy between historicity and theological truth is given life.

In fact, the reactionary reading of the text didn't stop the attacks on Christ's divinity, and so perhaps we should consider another reading. What if we instead read the text as revelatory and prophetic, proclaiming God's presence and faithfulness, even in the midst of trouble and sin? The image of the text as prophet is an apt one because it helps us remember that Scripture is not a simple chronicling of events. For example, when Christ teaches about a farmer sowing seeds in a field, he's not making an ontological point about a historical farmer, one with particularity, with locality; rather he is making a revelatory and prophetic point about the kingdom. In fact, when we look to the narratives surrounding Adam and Eve, they are shrouded in mysterious events: a talking snake, an enchanted tree, a woman is born out of a man's side. We might do well to ask questions concerning not what events took place in the garden but why the garden in the first place? Why the suffering and the fall from grace? If these events are a kind of moral narrative, to what do they point?

Yet we've come so often to hone our focus on the minutiae of the narrative—did Adam and Eve have belly buttons, what kind of snake was it, did they eat paleo or vegan? Instead, we might be better off focusing our attention on what's being obviously communicated from the Genesis account, exploring what God is revealing to us about his faithfulness and perfection rather than where the snake came from and why it's in the garden in the first place—that sort of work is best done by poets like Milton. One of the clearest takeaways from the Adam and Eve accounts in Genesis is the notion that God walked with the couple and still cared about them after the fall. Here, we see God making clothing for Adam and Eve and giving hedges of protection for Cain; and, though saddened by their state, God yet has paternal affection for them, the couple that represents humankind, as is clear in the narrative.

Along with the historicity of Adam and Eve, the passages dealing with the fall have also led to differing interpretations throughout Christian tradition. We find just such a range of differing interpretations in the works of Irenaeus and Augustine. Augustine did not view the creation narratives in Genesis to be literal; in fact, he published a whole paper cautioning against such a reading.[4] He viewed Adam and Eve to have been created perfect, to have fallen away from perfection because of sin, and he believed that they (humankind) would someday be made perfect again through Christ's work. In stark contrast, Irenaeus (second century AD), who lived only a few generations after Christ and who preceded Augustine, offered a drastically different reading of the Genesis creation narratives. In his *Against Heresies*, he argued that Adam and Eve are to be seen as spiritual infants who were

---

[4] See Augustine, *The Literal Meaning of Genesis*.

continuously in the process of growing into perfection.[5] In his view, though he agreed with the author of Genesis that Adam and Eve were created good, he didn't necessarily think the author meant that they were created *perfect*. For Irenaeus, perfection—which sometimes goes by the name of holiness or sanctification—was something that one had to grow into, not be born into. A privation of perfection was not a sin but a nascent state of existence.

Augustine and Irenaeus could not have differed more drastically on their views of Adam and Eve and the fall. And so here we are, in the twenty-first century, trying to understand early Genesis in light of contemporary science doing the same thing that they did in the early church. No one doubts the orthodoxy of either Augustine or Irenaeus for having differing opinions on Adam and Eve. In fact, the church has historically valued such voices to remind us to keep what's central in these texts central: that humanity has a severed relationship with God that needs fixing and, ultimately, perfecting.

THE PROBLEM OF PREDATION AND DEATH

As I write this, last night my daughter, who is seven, had severe abdominal pain and needed to be taken to the emergency room for possible appendicitis. Everything seems routine in life until it is not. When my wife texted me a picture of my sweet daughter with an IV in her arm ready to get a CT scan, I was overwhelmed by my concern for her. I kept thinking that this isn't normal. Kids shouldn't need hospitals, people shouldn't get gunned down, loved ones shouldn't die.

Let's face it. The forecast for our lives is cloudy with a likelihood of suffering. We're born, we suffer, and we die—there is no loophole. And what we know about evolutionary theory is that it has been this way for a long, long time. Well before humans were ever on the scene,

---

[5]For a good summation and explanation of *Against Heresies*, see Thomas Holsinger-Friesen, *Irenaeus and Genesis: A Study of Competition in Early Christian Hermeneutics* (University Park, PA: Eisenbrauns, 2009).

the organic life on this planet existed in a culture of violence, predominance, and predation. Every animal competed with every other animal to fight for scarce resources. This competition invariably left less fortunate animals without. Countless animals—including our closest prehuman ancestors—watched their offspring suffer and die before their eyes. As Lord Tennyson famously wrote, nature is "red in tooth and claw."[6]

And it wasn't just complex animals. Long before any animal was on earth, microorganisms were competing in the same way. Certainly, they didn't have the same complicated features as rather evolved animals, but they nevertheless suffered in their own way and eventually died. Death, it is clear, was built into this system long before humans were around. In fact, every beautiful creature you see, from the peacock to a newborn baby, depended on death for its own existence. And that's a hard pill to swallow. Death is not a consequence of sin but rather a *necessary* ingredient for life.[7] In the great chain of existence, if organisms don't die then life cannot spring forth. Think about it—I know it's a sobering thought—but someday you will die. No matter how much your body is preserved, the worms will get in. Your body will decompose and become food for a host of bugs. Birds will someday come and eat the bugs, then something eats the birds . . . you get it. It's not a fun thought, but it is absolutely how the world works. There's no getting around this one.

The place of predation and death in God's creation gives me more intellectual trouble than does the place of a historical Adam and Eve and a literal fall because I think there are theological explanations that are more reasonable to accept than the problems that are posed. The necessary place of predation and death in evolution is certainly, by all

---

[6] It is worth noting that Tennyson wrote this line before Darwin published *The Origin of Species*. Darwin was not responsible for creating the problem of natural evil.

[7] It should be noted that death is not technically a part of evolutionary theory. For evolution to work, one simply needs differential reproductive rates and heritable characteristics. Nothing *has to* die. Still, without death there would be massive overpopulation that would limit reproduction rates and by extension the evolution of organisms.

accounts, the most difficult reality for Christians to reconcile with Scripture and tradition. It seems clear that the shaping of this whole system—this whole universe—has always included death. And either God created the universe and had no idea that it would evolve this way—which we can all agree seems unlikely and certainly unorthodox—or God knew that any creation that would have the capacity to be free and evolve would necessitate predation, suffering, and death. And, frankly, at first glance this doesn't seem to line up with what Christian tradition tells us about the love of God. So this is, indeed, a problem for how we understand the Christian faith if we are to accept evolutionary theory.

So what does the presence of predation, suffering, and death in creation say about Jesus? I want to first acknowledge that I don't believe what I'm about to say is wholly satisfying. Furthermore, I have come to the acceptance that this problem of death might not be a puzzle that can be solved. Certainly, in my own Christian experience, I have had to live with the tension and the heartache that death has tangibly brought, not to mention the existential pain of the idea of death. Still, in the Christian faith, the idea that death is necessary for life is nothing new. In fact, across the entire Christian church we anticipate no church holiday more than the death and resurrection of Jesus Christ. Eastern Orthodox and Roman Catholic churches, for instance, hold midnight Easter services because they simply cannot wait another moment to celebrate the resurrection. Many Protestant churches have these celebrations at the crack of dawn. Even a nominal Christian will go to church on Easter. Why? Because, as I've heard cultural Christians typically say, "That's the important church day." Our whole Christian theology is focused on the death and resurrection of Christ.

I could go on and on about the countless books that have been written about theories of atonement, the Old Testament foreshadowing of the resurrection, or the fact that Christ is victor over death.

In fact, most Christian scholars agree that it is best to look at the death and resurrection of Jesus as one event rather than two separate events. We don't have resurrection without death. Christ's death was necessary for life.

To reiterate, while I do not believe this to be a wholly satisfying conclusion, I do wonder why I have more trouble with the notion of death being necessary for life from a biological perspective than I do from a religious perspective. While my Eastern Orthodox friends would say they don't *like* the idea of death, they would acknowledge that in some way all must die to experience life. My Roman Catholic friends have a similar theology. Perhaps, then, one of my tasks is not to find an explanation for death's existence in God's creation, but instead to articulate how the kingdom of God generates life out of death. In this way, we might come to focus on life rather than death.

PART TWO

UNDERSTANDING OUR SCIENTIFIC LENS

THE NUTS AND BOLTS OF EVOLUTION

A FRIENDLY DISCLAIMER: As this book will hopefully be read by educated laypeople and some scholars in the fields of theology and science, I was torn whether to include a chapter on evolutionary theory. Ultimately, if you, the reader, are interested in better understanding the basics of evolutionary theory, this chapter might be worthy of your attention. If you just finished a biology class at a local college, you might want to move on to the next chapter. My aim here is to make sure everyone is up to speed moving forward.[1]

Consequently, I would like to add a few notes of clarification concerning some common terminology within the field of evolution. It is easy to get tripped up when reading phrases about evolution that personify evolution. In other words, if you don't like it when the process

---

[1] It is not my intention to oversimplify this section about evolutionary theory. Yet I am sure that scholars in the field might make that claim. Because this book, and certainly this section, is intended to get laypeople up to speed on the theory, I hope that scholars in the field can forgive my use of some analogies that, if taken to their conclusions, don't sync with more complex evolutionary theory. I also want laypeople to know that some of these details are not conjecture and are scientific theory, but for some ideas listed in this section there is no established consensus. Here are two other theories and monograph-length arguments for natural selection: Andreas Wagner, *Arrival of the Fittest: Solving Evolution's Greatest Puzzle* (New York: Penguin, 2014); and David Quammen, *The Tangled Tree: A Radical New History of Life* (New York: Simon & Schuster, 2018).

of evolution seems to have a sense of agency, then you're in good company. Neither do I. Often when people within the field talk of evolution, they use phrases like "evolution selected this trait" or "this trait was added by evolution." Let me be clear: evolution is a passive process. It has neither will nor agency. Evolution has no volition or ability to be any kind of first causation. I'm not sure where this habit got started. Perhaps using phrases like this became a shorthand way of articulating larger, more complicated concepts. For example, someone might say, "Evolution provided humans with a large brain," instead of saying, "The process of evolution by natural selection happened to culminate in humans having a larger brain than all other animals." The words *process* and *happened* are quite important words—they are signifiers that evolution is not some kind of metaphysical force. Thus, if you read anything in this book or hear such language from others in the field that assigns evolution any sort of agency, please note that (while I try to avoid such usage in my writing), it is common parlance in the field. No serious philosopher, theologian, biologist, or sociobiologist considers evolution to be a metaphysical process.

Another term we ought to discuss briefly is the word *theory*. It should be noted that in scientific terms the word *theory* differs from how we use it in everyday language. Theory for nonscientists is an idea that expresses possibilities more than it expresses probabilities. We might use language like, "In theory, this car shouldn't be sounding like this!" or "Communism works in theory but not in practice," or even, "In theory I would love to take a vacation, but I just can't afford the time off." This colloquial use of the word works against scientists who want to communicate complicated science to laypeople who aren't in the trenches of empirical data.

When scientists employ *theory*, they mean quite the opposite of our colloquial usages. In science the word *theory* expresses an empirically tested set of ideas. Some theories, through scientific history, have had a short life: they are born, are tested, and then die. Others, most notably the theory of general relativity (gravity) and the theory of evolution have undergone countless tests and are still alive and well. Some of our confusion likely comes about because we don't often parse out the differences between a *law* and a *theory*. Simply put, a scientific law predicts *what* happens, and theory tells us *why* it happens. A theory will never be so proven that it is finally classified a law, just as photosynthesis will never change into an apple. One is the process, the other is the result. In other words, a theory is not a stepping-stone to a law. Their relationship is not ordinal but hierarchical, for they have different functions.

The crucible that the theory of evolution by natural selection has undergone is truly remarkable—with a wide range of independent scientific disciplines reaching similar conclusions. Numerous fields, such as anthropology, geology, biology, genetics, astronomy, sociobiology, chemistry, and astrophysics have all come to the conclusion that the universe is incredibly old. Now, it should be said that the age of the universe in, say, physics, geology, and astronomy, doesn't *necessarily* mean that organic evolution takes place, but it does prove to be strong support for biological evolution. One of the virtues I so appreciate about science is that there is nothing scientists love more than for new evidence to blow up an old theory. And given that such disparate fields as those just listed confirm (in one way or another) the theory of evolution, we should at the very least be willing to consider what they have to say.

As I mentioned in chapter one, my goal in this book is not solely to convince people about the veracity of the theory of evolution but rather to show how seeking an understanding of it is beneficial to our Christian faith and our journey toward a holy life. In order to get to that payoff, I believe it is necessary to address typical fears of natural selection. Fear of a thing is poor justification for not seeking to understand it. I hope these ideas on basic evolution serve to give us both a common understanding and a unified terminology. If we are to dialogue about this very important topic, we must first learn to share a common understanding of process and speak the same language;

by doing so, we can work to avoid frustrating misunderstandings and unintentional slights.

Furthermore, I think it would be a shame to confuse good reason with certainty. What I mean is this: just because we don't know *everything* about how life was created doesn't mean we throw away the whole idea. As an example, we don't know everything about Abraham Lincoln and why he performed certain actions; and we don't even know how precise historical events unfolded. But that doesn't mean we throw away the whole idea of Abraham Lincoln. I could even argue that we aren't certain of all the circumstances in the Bible, from how many Israelites actually left Egypt to which thieves on the cross denied Jesus—one or both. This lack of knowledge doesn't embolden us to throw out our faith. To say that statements like, "You weren't there when the world was created, so how do you *know* it was created that way" are not only unhelpful, they're unreasoned.[2] With that logic, we can't articulate the veracity of any biblical events or even the existence of our great-great-grandparents.

More importantly, there's no reason why God couldn't have started this process of evolution 3.7 billion years ago when organic life started on earth. I find it more impressive that God could create a creation that could *itself* be a creator. It's why we find organic life more impressive than inorganic matter—puppies are more interesting than rocks. Why? Not just because they are cute—they certainly have that going for them—but also because they can independently bring forth life. That, by all accounts, shows the glory of God.

WHAT IF SUPERHEROES COULD REPRODUCE?

When I was a kid, I dreamed of being a superhero. I watched a lot of *X-Men*—my parents would probably say *too much X-Men*. But I think

---

[2] See Ken Ham's repeated statement, "Were you there?" during his debate with Bill Nye. "Bill Nye Debates Ken Ham," *Answers in Genesis*, February 4, 2014, www.youtube.com/watch?v=z6kgvhG3AkI.

the reason that so many kids are enthralled with superheroes is that we all want to have some kind of special ability—a special power that sets us apart. I was no different. My favorite superheroes were always the ones who looked completely normal but had one or two special abilities. It's slightly embarrassing for me to even write this, but I often thought that if *they* only had one or two superhuman traits, then maybe I might as well. However, it was only a matter of time before I came to realize that I wouldn't be shooting lasers out of my eyes after all—I just wasn't that special. But after I began studying the theory of evolution, I came to realize that every succeeding generation of a particular species has become, in its own way, a mutant superhero.

So what do superheroes have to do with evolution? Imagine with me if superheroes had children together. And when those children grew up, they had children. What if the superhero's special traits could be mixed and passed down through the generations? If they could, every subsequent generation would become stronger and stronger. Think of the amazing children that Superman and Wonder Woman would have—the strength, the speed, the intelligence! But here's the exciting thing about evolution in the real world: we *are* these reproducing superheroes, even if we don't necessarily realize it.

If one of our ancestral species that existed hundreds of thousands of years ago could peer into our contemporary time they would, no doubt, believe us to be superheroes. They would notice many features that are different from theirs, perhaps the greatest of which would be our advanced brains. For *Homo sapiens*, our brain is our superpower. Our highly evolved brains give us the ability to use complicated language and communication skills, the ability to create complex tools, and the ability to construct modes of transportation and dwelling spaces that are unprecedented in the history of organic life. In order to understand how impressive humans really are, we need to first grasp the two main concepts of evolution: *mutation* and *replication*. As in the case with superheroes, mutation in evolutionary terms expresses the notion of altered or mutated traits from the previous generation. Replication simply refers to the process of passing on advantageous traits to one's kin.

In the animal world, mutation and replication happen by and large through the same process: sex.[3] An organism exchanges DNA with its partner and reproduces something that is, in one creature, the product of both partners, though identical to neither. Sexual reproduction is extremely important to evolving complex organic life. When I was a kid I would go to my grandparent's house on my mom's side of the family and my Grandma Hartman would grab my cheek and tell me how I looked exactly like a Hartman. "Oh!" she would say, "Matt's the spitting image of Grandpa Don! He's a Hartman through and through!" The next weekend I'd go to my other grandparents and endure the same treatment. My grandpa would slap me on the back and proudly shout, "Look at this little Hill! If he's not a full-blooded Hill then I don't know who is!" But I wasn't either of those things. I was a mutated "Hartman-Hill"—neither fully Hartman nor fully Hill.[4]

If you have siblings, as I do, you'll look at your siblings and see strong traits in them that favor one side or the other. But even if you have siblings who bear striking similarities with you, you still look different from each other. You have the same parents and may even be the same gender as your sibling, but you are your own person, a unique creature. That's because you are, in a sense, a mutant—derived from a unique combination of genes and environmental influences—and so are your siblings. When two organisms sexually reproduce, they set up a kind of genetic lottery. Sometimes the traits passed on to the next generation are helpful, and sometimes they aren't.

---

[3]In an effort to keep the biological language simple, I am using the word *replication* here. Technically, the correct word would be *selection*—as I'll expound on in this and later chapters. But I want readers to know that I don't mean to mislead them and that I am also aware of the difference between replication and selection.

[4]This wording of mutation is oversimplified as well. While there are sixty changes that take place from parent to child, the alterations are out of 3.5 billion code letters. Thanks to Darrel Falk for pointing this out.

There are these two X-Men characters, Gambit and Rogue, who have a romantic relationship. The problem is, Rogue has a mutant power that doesn't allow her to touch another living thing without killing it. Their relationship, as one might expect, doesn't exist without a bit of tension. Rogue's mutant power might help her fight bad guys, but it's not very helpful for reproduction. She might really want a family with, say, Gambit, but she'll never get it because she has an unhelpful mutation. There will be no little Rogues running around, and, in effect, her mutant power will die with her. Sexual reproduction is so important because it allows good traits to be passed on to the next generation—what is often called "being selected"—while bad traits that don't help one's reproductive fitness either die off or diminish their influence over generations. Essentially, every generation, on the whole, is more fine-tuned for its current environment than the previous one.[5]

Over eons of time sexual reproduction has also kept us healthier. As many of us palpably know, *Homo sapiens*, as well as all other animals, are in a kind of biological arms race with viruses and bacteria that are trying to survive, and in doing so they can kill us.[6] More precisely, they're just trying to stay alive, but these organisms nonetheless cause great harm and even death to our species (e.g., AIDS, Ebola, influenza). Hypothetically, what if we reproduced asexually so every person in the world was a genetic clone of the next? If a virus started wreaking havoc, and we had no vaccine, there would be little in the way of stopping it. The recent Ebola virus outbreak in West Africa is a great example of why sexual reproduction protects humans from quick extinction: some West Africans are actually biologically immune to the Ebola virus because of the way their DNA is mixed from their parents.[7] Picture for a moment that the Ebola virus ran

---

[5]Both Charles Darwin and Robert Malthus observed that more offspring were typically born in a particular species than could survive. The ones that survived had better chances of reproducing. Thus, their genes were carried to the subsequent generation.
[6]For more info on this "biological arms race," see Matt Ridley, *The Red Queen: Sex and the Evolution of Human Nature* (New York: Perennial, 2003).
[7]Douglas G. McNeil Jr., "Many in West Africa May Be Immune to Ebola Virus," *New York Times*, September 5, 2014, nytimes.com/2014/09/06/health/ebola-immunity.html.

rampant and wiped out most of the human race. The groups of people in West Africa who were immune would be the ones to produce the next generation of humans. Their mutation not only kept them safe from the virus, but it will also protect their kin in whom the mutation has been replicated.

It's important to note, however, that mutation and replication are *not* random, as is often popularly believed. It's foolish to perpetuate the myth that organisms evolve by chance. While mutations are arbitrary, or sometimes even called *blind*, and are beyond the organism's control, they are birthed and tested in a milieu that is anything but random. There are very distinct circumstances that determine whether or not a mutated trait is beneficial or detrimental to the organism—circumstances such as harsh environments, a lack or abundance of predators, scarcity of food, and so forth. All of these factors are very particular in their degree and essentially put the mutation through a kind of sieve. So one cannot say, "All animals evolved by chance," without some *major* caveats to the nonrandom circumstances that seemingly random mutations find themselves in.[8]

EVOLUTION AS FLIPBOOK ANIMATION

I was showing my kids *The Magic School Bus* cartoons—the one where Ms. Frizzle takes a fanciful bus to explore crazy destinations like space or the human body. We had finished watching the one where the bus shrinks and travels through the human digestive system, and my three-year-old daughter completely lost it.

"What's wrong, honey? Why are you crying?" I asked.

"Is there a school bus inside me right now?" my poor little daughter said through tears.

"Oh honey, no. That's just a cartoon."

"But will it hurt when I poop out the school bus?"

---

[8]For some quick and fun lessons on the theory of evolution, see the Crash Course YouTube channel (youtube.com/user/crashcourse). Some of their anecdotes were helpful when I compiled this chapter.

THE NUTS AND BOLTS OF EVOLUTION

That was a good question! I bet *The Magic School Bus* franchise never thought they were producing a horror movie! I proceeded to explain to her how a cartoon is made and how she didn't need to be afraid because it wasn't real. To do this we put together a little flipbook cartoon out of paper. On several pieces of paper I drew a stick figure throwing a ball in various stages of the exercise. Then we flipped the little book over and over again until she saw that cartoons were just a series of pictures put together in rapid succession.

I know this sounds silly, but hear me out. Evolution is like flipbook animation.[9] If you string together enough pictures and look at them in rapid sequence, then evolution *appears* completely fluid, when in fact the process of evolution is actually numerous solitary events.

Sometimes, when I talk to people in my church about evolution, they get tripped up on macroevolution but not microevolution.[10] Put simply, microevolution is small, minor changes that take place within a species in short periods of time. We see microevolution a lot in our world today, so it's no surprise that people aren't bothered by it. Take for instance the peppered moth found in England. Before the industrial revolution, the moth was primarily white and peppered with black dots—with some outlier moths mutated as fully black in color. This largely white color helped the moth blend in with the lichen that it was landing on. After the Industrial Revolution of England killed off a certain kind of white lichen and blackened the trees, the moths that naturally were in the minority and darker in color began to thrive, while the whiter moths started to die off. In evolutionary terms we would say that the darker pepper moth was *selected for* in the fiercely competitive environment. The species didn't get up one day and

---

[9] While this idea first came to me from various lectures by Richard Dawkins, the analogy of explaining evolution as a flipbook is a common one and certainly not original with me. See this helpful flipbook video, "550 Million Years of Human Evolution Flipbook" on YouTube (youtube.com/watch?v=Epwer3w96ew).

[10] To be clear, macroevolution, per se, is not actually a concept in evolutionary theory, notably because "micro is macro," as I explain in this section.

decide to become darker; rather, with every succeeding generation, there happened to be more dark moths around for mating.

Another, scarier example of microevolution is taking place right in front of us every time we take an antibiotic. As many of us know, many of the bacteria that make us sick, from strep throat to MRSA, are increasingly becoming antibiotic resistant. Not only does this tend to make it harder for us to get better quickly, but also we have to keep taking stronger and stronger doses to just stay on top of the sickness. There's no better example of this than a recent Harvard study involving the bacteria *E. coli*. Scientists from Harvard's Kishony Lab put together a supersized petri dish that was two feet by four feet and had various chambers mostly walled off. Each section of the dish had increasing amounts of antibiotics. One section had just a little more than would kill *E. coli*; the next had ten times the original amount, then one hundred times, then one thousand times. At first, the bacteria were locked into a spot where they could easily spread. But then a mutant *E. coli* emerged, by chance, and had the space to expand and compete with other antibiotic-resistant mutants. This pattern continued until the whole petri dish—even the section with one thousand strength antibiotic—was overtaken with a mutant strain of *E. coli*. Even though *E. coli* reproduces at a much faster rate than humans, the scary part of this study is that it only took only eleven days for the bacteria to be one thousand times stronger than it was when the experiment started.

Microevolution is easy to believe in not just because our life depends on it, as is the case with understanding *E. coli*, but also because we can see it happening before our eyes every day. Every succeeding generation of a family looks a bit different from the previous—for good and bad! Every pet that we breed, chicken that we eat, or fruit that we pick is a small example of microevolution.

Belief in macroevolution is sometimes more difficult for people. Macroevolution involves major evolutionary change usually over long periods of time that is harder to see today without digging deeper into

THE NUTS AND BOLTS OF EVOLUTION

fossil records. While there are numerous examples from fossil records that document macroevolution—from the manatee losing its hind legs to the long history of human descent—I often hear people say something like, "I believe in microevolution, but can't make the jump to macroevolution." I get it. It can be scary, or it might seem that the evidence isn't quite there for macroevolution. But perhaps we would fear the evidence less if we understood that macroevolution *is* microevolution, just over long periods of time.

It might help to visualize the entire evolutionary history of organic life as a gigantic flipbook animation, with each page representing a century in the life of this planet. Now this book would be massive and almost unimaginable in size (about 40 million pages), and it would contain every generation of every species of life. If I take the time to tediously flip page by page and show the small changes from generation to generation, it wouldn't ruffle any feathers of those comfortable with microevolution. The page-by-page flipping is the same activity that I did at my grandmother's home when we would look at old photo albums together: "This is your great-great-grandfather, and you have his smile." No problem there. But I remember one time when my Grandma Hartman grabbed a bunch of pages and flipped way back. She pointed to a man, "BigJud," from Civil War time. Grandma Hartman said I was related to BigJud, and I believe her, but I don't look anything like him.

So what if instead of looking at one big flipbook of evolution, we looked at two flipbooks of two particular animals side by side? And instead of flipping page by page in our flipbooks, we started to grab and flip between chunks of pages, whole chapters even? That might look pretty similar to macroevolution. We can grab a chunk of pages and flip from today, where manatees and elephants are distant relatives, to sixty million years ago, when they shared the same four-legged and hoofed land-dwelling ancestor that lived near shorelines. And then, what would happen if we did the tedious work of turning one page at a time simultaneously in both books? At some point we would find that elephants and manatees looked very different from each other. If microevolution is page by page, then macroevolution could be seen as chapter by chapter—a collection of microevolution events flipped all at once. Perhaps if we did such, we would be less surprised that the elephant and manatee look different today.

It's common to wonder that if macroevolution is in essence microevolution but on a larger scale, then at what point does one species become another species? The answer to that question might best be illustrated with my love for my wife. Twenty years ago I did *not* love my wife. I didn't love her because I didn't know she existed; we grew up on different continents and hadn't yet met. Today, I'm deeply in love with my wife, and we've been married for nearly fourteen years. And though I *am* in love with my wife, but *wasn't* twenty years ago, I would be hard-pressed to tell you when exactly I actually fell in love with her. The process was incremental and gradual. We never had a single moment where we professed our love for each other, marking that moment by tattooing it on our arms. And even if we did have such a moment, that love would be an anemic version of the love I have for her now. So though I can't point to when we fell in love, I know without a doubt that we are.

The same thing is true of my personal relationship with Jesus. I grew up in the church and had various moments that were sacramental and important to my faith, like baptism, Communion, and inviting Christ to lead my life. But when *exactly* did I become a Christian? Others have a powerful conversion story illustrated by radical changes in behavior. Not me. For the most part I grew up loving Jesus, and I still do today. At *some point* I must have made my faith my own—not living off my parents' faith—but I truly don't know when that was. I just know that while I probably didn't have any real relationship with Jesus when I was a baby, I very much love Jesus now, and that love is personal.

What I'm trying to suggest is that there are other areas in life where we can appreciate the power of a moment as well as the growth of something over time. So it is with evolution. Macroevolution is a series of microevolution events, and that would suggest it's all process. But at some point—even if we aren't fully aware of when or even how it happened—the manatee and the elephant broke off from a common ancestor and became different species.

WHAT MAKES HUMANS SO SPECIAL?

Humans are special. There are no two ways about it. There are theological notions that give testament to our special nature that I will discuss in detail in the third section of this book—but I will quickly list them here: (1) We are endowed with the *imago Dei* or the image of God, and (2) God could have become incarnate as any organism, but he chose to become incarnate as a human. It is an understatement, indeed, to say that these divine interactions make us special.

To be clear, humans are special for more than theological reasons. We are the only animal that lives on all seven continents, we have lived in space, on the moon, and sometime soon we will likely live on Mars. We have communication networks that are unparalleled among organic life, wildly fast transportation technology, and medical advances that dwarf all other animals' attempts at self-healing. We are impressive and unrivaled. And one of the biggest natural reasons for such innovation is our highly evolved and advanced brain.

Besides enabling us to be empathetic and to have prosocial tendencies, our brains have allowed us to do something that almost—*almost*—takes us out of the evolutionary system entirely: we can defy our genes by ignoring certain genetic inclinations.[11] This defiance manifests itself in various ways. For instance, our bodies are hardwired for

---

[11]For those of you who are interested in a more technical explanation of how Christians might be positioned well to start overcoming genetic inclinations, see Matthew Nelson Hill, *Evolution and Holiness: Sociobiology, Altruism, and the Quest for Wesleyan Perfection*, Strategic Initiatives in Evangelical Theology (Downers Grove, IL: IVP Academic, 2016).

sex—it is pleasurable and has been the process by which organisms have been reproducing for eons. Every one of us is a descendant of countless generations of relatives who procreated. But humankind, with our advanced brains, has purposely developed a way to experience the pleasure of sex without having children! At first glance, this development might seem obvious: we desired the pleasure of sex without the responsibility that comes with having children. The profundity of such a reality is that the practice of contraception works directly against the genes we've inherited—genes that urge us to procreate. Yet we humans have evolved to a place where we don't have to obey all of our genetic proclivities. We are not slaves to them. We can make free choices.

We see the same freedom expressed in other practices often found in religious life, from fasting to celibacy. Certainly, it helps that many humans have an abundance of food, making fasting less dangerous and risky. But to deny oneself food, for some higher purpose that is nonphysical, is unprecedented in the animal world. Celibate priests, nuns, and new monastics also clearly demonstrate the way our brains have allowed us to take the driver's seat not only in our lives but also in evolution.

The way that humans care for the infirm is also absolutely remarkable and goes against selfish traits from our evolutionary past.[12] As a species we purposefully keep people alive to reproductive age even if a genetic mutation isn't helpful. Type 1 diabetes is a good example of this; while this disease isn't directly inherited, those who have an immediate family member (e.g., a sibling) with the disease are six times more likely develop the illness. But humans, with the help of our advanced brains, were able to come up with a workaround: we created artificially produced insulin. In 1921 we developed the ability to give insulin to children with Type 1 diabetes and extended their lives to decades. Now, these same children live fully into old age and reproduce with the rest

---

[12] It should be noted that much research points to biological and sociobiological roots for altruism. As we'll discuss later in this book, part of the task for Christians is to nurture said natural urges.

of the population. While this achievement is noble and virtuous, it's not particularly helpful from an evolutionary perspective.

I said earlier that humans are *almost* outside of evolution because that statement demands a major qualifier. Yes, our brains give us the ability to purposefully go against our gene fitness, and that very much undermines the process of evolution. But we are still evolving, just in ways that are more difficult to predict and notice. For instance, our spike in gluten and dairy allergies should not be surprising to us. Humans have only been eating refined wheat and milk products for about ten thousand years. That's a mere blip in the scope of human evolution. Of course, many of us can't tolerate these kinds of nutrients—some of us simply aren't designed to eat them. With the rise of agriculture came the rise of technology—and because it took fewer people to make more food, this left others to invent and discover. This growth in technology, by and large, helps us get to reproductive age, despite maladaptive side effects. In other words, there wasn't enough time for those who couldn't tolerate wheat and dairy to die off. So yes, we are still evolving, but our brains invent so fast and with such reach that it's difficult to predict what parts of us are helpful or harmful from a reproductive fitness perspective.

OUR STORY: THE PAGES OF EVOLUTION

Now that we've discussed some of the basic concepts of evolutionary theory, I'd like to spend some time looking at how the history of organic life unfolded on earth. Since many narratives of evolution exist—some more truthful than others—these concepts are particularly important for those of you who desire a brief overview of the story of evolution's history from a Christian perspective.[13]

---

[13] Here are two good resources for those who want to better connect their faith with evolution and science: Darrel R. Falk, *Coming to Peace with Science: Bridging the Worlds Between Faith and Biology* (Downers Grove, IL: InterVarsity Press, 2004); and Kathryn Applegate, *How I Changed My Mind About Evolution: Evangelicals Reflect on Faith and Science* (Downers Grove, IL: IVP Academic, 2016).

As you might expect, there is some discussion as to how organic life got started 3.8 to 4 billion years ago. All organic life comes from simple beginnings, to be sure, with the most likely theory pointing toward hot undersea vents where basic molecules such as amino acids and ribonucleic acids (RNA) could be assembled. Some even hypothesize that life on earth started with a "panspermia" event—where organic life somewhere else in our solar system "seeded" earth via asteroid impacts. While this theory isn't widely accepted, it illustrates an important point: scientists are confident, but not *completely* sure, about how life got started on earth.[14]

In 1828, German chemist Friedrich Wohler used inorganic chemicals to synthesize organic chemicals. This experiment showed that inorganic and organic chemicals follow the same laws and that life could, possibly, emerge from nonlife, or more particularly, how inorganic material could be made organic. And while organic evolution has been occurring for 3.8 billion years, for 2.1 billion of those years all that existed were simple single-celled organism called "prokaryotes." Around 1.7 billion years ago, some prokaryotes evolved through natural selection—mutation and replication—by developing a partnership with an archaeal cell, helping to form into larger single-celled organisms called "eukaryotes." So, in a very real way, when we look at the first few flipbook pages of organic life, we can observe our early relatives, the prokaryotes, single-celled organisms that lived in steamy gas communities and ate chemicals. A humble beginning, to be sure. Yet think about the beauty of life—how, even in their single-celled simplicity, these little life forms are wildly complex compared to the inanimate stardust they evolved from. It's truly remarkable, and it very explicitly reflects the majesty of the Creator.

---

[14]While there are numerous books and videos available on the subject of evolution, I highly recommend Karl Giberson and Francis S. Collins, *The Language of Science and Faith: Straight Answers to Genuine Questions* (Downers Grove, IL: InterVarsity Press, 2011). I also recommend the CrashCourse series on YouTube. It's accessible, accurate, and entertaining, and some of my explanations in this chapter are indebted to it. For a resource that discusses the natural limits to science, see Wendell Berry, *Life Is a Miracle: An Essay Against Modern Superstition* (Washington, DC: Counterpoint, 2000).

NICHE ENVIRONMENTS

Even though this first phase of evolution spanned an epoch of billions of years, life continued to evolve as these prokaryotes moved closer to the surface of the water and began the process of photosynthesis by using the sun's light to get energy, and by doing so, helping convert carbon dioxide from the air into oxygen. Back then, over two billion years ago, the oxygenation of the air was deadly to the abundant anaerobic organisms. Dramatic atmospheric change, brought forth by photosynthesis, initiated the first example we have on earth of mass extinction: the Great Oxygenation Event.

I want to pause the timeline of evolutionary history by discussing an important idea brought forth most famously by American paleontologists Niles Eldredge and Stephen Jay Gould. Punctuated equilibrium expresses the notion that evolution does not necessarily take place with a steady flow of even change. Rather, evolution often occurs with dramatic bursts (punctuation), followed by periods of little development—or stasis—that bring about equilibrium. In other words, there are certain climates that have the right ingredients to bring about high quantities of change.[15]

I'm a fly fisherman, and one of my favorite parts of fishing isn't necessarily catching fish; it's watching the change in the environment. I remember fly fishing with my friend Mark, who wanted to show me the trico (or *Tricorythodes*) mayfly hatch—when the trout feed on the bugs at certain times that correlate to the conditions of their environment. Mark had been fishing the trico hatch for years and had timed it with his watch—literally. It was 10:00 a.m. and we were standing in the cold Oatka Creek water watching all the tricos fly in their mating flights over our heads. Neither of us had our lines in the water; we just sat quietly and watched. Mark told me that at 10:10 a.m. the creek would light up with fish emerging in every direction as the tricos started to fall from

---

[15] Punctuated equilibrium is not, of course, without its critics. See D. C. Dennett, *Darwin's Dangerous Idea: Evolution and the Meanings of Life* (New York: Simon & Schuster, 1995).

the sky after mating. Sure enough, at 10:10 a.m. (on the dot!) we saw just a few trico fall to the surface of the water, knocking other mayflies down with them. This started a chain event that turned the previously calm creek into a foaming rapid with the number of fish that were jumping for falling flies. I was shocked. I didn't know it at the time, but the environment was quietly teetering on the edge of change until something pushed it into commotion, and in dramatic fashion. The fish —biologically evolved to capitalize on the change in the environment— were perfectly positioned to take advantage of their surroundings.

In organic evolution, extinctions are tipping-point events that dramatically alter the environment and accelerate the process of evolution. If we experienced a major event on earth—for example, a meteorite hitting the earth, a deadly virus wiping out 90 percent of organic life on earth—then there would be room for new sets of organisms to emerge. For our own story most biologists agree that if dinosaurs hadn't gone extinct in quick fashion, mammals—and ultimately humans—would never have had the space to evolve in the way we did. When mammals and dinosaurs coexisted, a mammal's primary job was to hide from its reptilian predators so that it didn't become their lunch.

The idea that mass extinction often speeds up major evolutionary development is a bitter pill to swallow. It appears that death is a necessary part of bringing forth life. In fact, I think this concept is one of the most difficult for Christians to reconcile with evolution. I will pointedly address this concern in section two of this book—but for now a word of warning is in line: any answer to the problem of death and predation is not an easy one.

There were five main extinction events that led to periods of punctuated evolution, where rapid change was commonplace, and where the extinctions helped to reshape the landscape of organic life on earth. Before we look at the five major events, there is one catastrophe that produced complex organic life in such a dramatic fashion that it's typically not considered part of the main five. The Great Oxygenation Event (occurring around 2.3 billion years ago) is sometimes called the Oxygen Catastrophe and stands out among all extinctions in earth's history because, by it, organic life first experienced rapid reproduction. This cataclysmic incident was the first major die-off, killing untold quantities of microorganisms that couldn't handle the dramatic increase in oxygen. But it did something else as well: it created what many in the field call "niche environments." With the addition of primitively evolved photosynthesizers, an influx of oxygen burst into the atmosphere in a relatively short period of time. This flood of oxygen changed earth's atmosphere, blocking some of the harsh radiations that prevented complex life from evolving. Not only had organic life found an endless supply of food—the sun—but also it had developed a niche that, while killing off an untold number of anaerobic species, allowed for the evolution of complicated life forms.

It wasn't until nearly two billion years later that the Cambrian explosion ensued (occurring around 540 million years ago), during which the earth saw the emergence of its first real complex organisms. Reflect a moment on the implications here: it took several billion years of evolution to get earth to a place where, in our eyes, it would be recognizable. But as we will see, complex life begets complex life, and punctuated evolution ignited by relatively rapid extinction would come fast and furious.

For our purposes I will only touch briefly here on the other five major extinctions that occurred after the Great Oxygenation Event: (1) the Ordovician-Sulurian extinction event (occurring around 450–440 million years ago), during which the earth lost about 60–70 percent of all species; (2) the Late Devonian extinction (occurring around 375–360 million years ago) during which 70 percent of all species became extinct; (3) the Permian-Triassic extinction event (occurring around 252 million years ago), triggered by a multitude of catastrophes that hit simultaneously—for example, asteroids, noxious floods, and a change in earth's atmosphere; this is earth's largest extinction, where about 90–96 percent of all species on earth perished; all organic life currently on earth descends from this 5 percent of remaining organisms; (4) the Triassic-Jurassic extinction event (occurring about 201.3 million years ago), during which 70–75 percent of all species became extinct; (5) the Cretaceous-Paleogene extinction event (occurring around 66 million years ago) where three-fourths of plants and animals became extinct. This last event is most notable for killing off the rest of the dinosaurs that weren't killed in the Triassic-Jurassic extinction. The likely cause of this calamity was a combination of volcanic eruptions that changed the atmosphere along with an asteroid that was about 110 miles across and struck the earth near the Yucatan Peninsula—ultimately reshaping earth's atmosphere for a time and wiping out the dinosaurs.

THE NITTY-GRITTY DETAILS: HOW COMPLICATED LIFE CONTINUED TO EVOLVE

We previously looked at how sexual reproduction became an important catalyst for fast evolutionary change as it accelerated the ways organisms mutated. If we turn back to 1.7 billion years ago, we'll notice that the eukaryotes—the larger single-celled organisms—developed organ-like features and began to reproduce sexually. Numerous single-celled eukaryote organisms started to work together in what is called "symbiosis," where a reciprocal relationship formed. Some of these relationships became so intertwined that one organism couldn't survive without the other. Triggered by numerous extinction events, the harsher the conditions, the more single-celled organisms were forced to work together in symbiotic relationships that resulted in more and more complex organisms.

It is also important to note a unity that is shared by all living things on earth, from amoebas to plants to buffalos to humans: deoxyribonucleic and ribonucleic acid, or more commonly, DNA and RNA.

THE NUTS AND BOLTS OF EVOLUTION

DNA is a double-stranded coil of information that consists of pairs of chemical groups called "bases." RNA acts as a messenger delivering instructions from the DNA to proteins to essentially carry out the plan-information originating in the DNA. Think of DNA as a roadmap and RNA as a courier delivering the directions to the proteins, the workers who actually get the job done. Because *all* life on earth has DNA, we can know that on some level all organic life is related—a concept that we'll discuss later. Life's common DNA patterns are part of the reason why we humans share around 98 percent of our DNA with the great apes (depending on the species) but also the same reason that we famously—or infamously—share 50 percent of our DNA with bananas. The key factor is not necessarily how much of the DNA is *shared* but how that shared DNA is *arranged*. I had a professor explain it to me this way: Hitler's *Mein Kampf* and Marx and Engels's *Communist Manifesto* both contain the same German letters; but how those letters are arranged makes all the difference in the world.

DNA replication, while typically copying a living organism with uncanny exactness, sometimes comes up with an error—or mutation. While many errors are simply innocuous, making the majority of mutations neutral—like the color of one's hair—DNA mutations can also range from being detrimental to the organism to being quite helpful. If the mutation is detrimental to the organism, it works against them passing on their genes, often contributing to premature death before reproduction. If the DNA mutation happens to be helpful to the organism—ushering them to reproductive age—the mutation becomes part of the new genetic code and is passed on through DNA-RNA replication.

As you might expect, this kind of systematic replication and mutation gave consistency and framework for life to evolve and produce a portrait of sustainable diversity. The world evolved to a place, after eons of slow, methodical change, where an abundance of life was poised to spring forth onto land.

At the end of the Cambrian period, during the Ordovician period, (roughly 485–445 million years ago) the organisms that thrived owing to photosynthesis began to make their way to costal lands and emerged out of the water, first at the coasts and then working their way toward the center of the land mass, becoming more and more complex and varied along the way. This activity was the beginning of life on the two supercontinents, Gondwana and Laurasia, that would later merge together as Pangaea, and is the primary reason that life on earth looks relatively similar no matter where you go. And while the seas were bursting with abundant and wildly diverse life, their niche was getting crowded, prompting organisms to move to land for less competition. After a while, the supercontinent started to team with life, just as earth's great oceans did, and life on land became more and more complex, leaving an opening for mammals—such as *Homo sapiens*—to stake out some territory.

THE OPENING FOR HUMANS

I am always troubled when I think of how mammals—and ultimately humankind—emerged out of the ashes of the last great extinction event: the Cretaceous-Paleogene catastrophe around sixty-six million years ago. Earth didn't just lose the dinosaurs, it lost about 75 percent of its flora and fauna. This great death toll created the niche space for mammals to rise to dominance. I will also address the tension between both life's dependency on death (the dependence of human existence on these great extinction events) and the theological truth that death is still, quite palpably, the great enemy.

If you look back far enough at the flipbook of humankind's ancestry, you'll find a funny little ancestor of ours: a tiny mouse-like mammal from which all humans ultimately descend. While the larger animals like the dinosaurs were rapidly becoming extinct due to extreme climate change and scarcity of food, this little creature wasn't particularly put out. It could burrow in the ground and wait out the tough spells of extreme weather, and it was small enough that it didn't need an abundance of food in order to survive—especially in contrast to the dinosaurs. Before the meteorite that hit the Yucatan Peninsula, these little guys were tasty appetizers for bigger predators. To oversimplify it, after the catastrophic extinction event, the small mammal's size proved to be a great advantage to the new niche environment.

If we fast-forward to around seven to eight million years ago, the human line of descent split off from a common ancestor shared by African chimpanzees.[16] It would be prudent here to quickly dispel a common misconception in evolution: we did not evolve from the monkeys we see today, nor are we a more "highly evolved" chimpanzee. Instead, we are more like cousins to them, sharing the same ancestors several million years ago. During this time humankind's direct ancestors went through various phases of evolutionary development, ranging from the transition to bipedalism (around four million years ago)—the ability to solely walk on two legs and freeing up hands for tools—to 1.9–1.5 million years ago *Homo ergaster erectus* who developed a variety of axe heads, displaying the first sign of technological advancements rather than mimicking past behavior—essentially, they learned how to improve technology, not just mimic it.[17]

This evolutionary advancement is critical for understanding human development because we all are able to build off of past discoveries and improve on previous technologies. We call this development "collective learning."[18] For example, while most of us have no idea how a car works, we can both learn how to use a car and also possibly discover small improvements that make cars run better. We likewise see collective learning at play in our work with computers: I don't need to fully know how processors work to be able to code cleaner, smoother programs.

---

[16]Nick Patterson et al., "Genetic Evidence for Complex Speciation of Humans and Chimpanzees," *Nature* 441, no. 7097 (2006).

[17]Michael Ruse and Joseph Travis, eds., *Evolution: The First Four Billion Years* (Cambridge, MA: Belknap Press, 2009), 270.

[18]David M. Carballo, *Cooperation and Collective Action Archaeological Perspectives* (Boulder: University Press of Colorado, 2013), 37.

Besides the development of agriculture, nothing has helped humans collectively learn more efficiently than the capacity for complex language. Most scholars think that one of the mutations that helped *Homo sapiens* edge past their competitors, the Neanderthals of Europe and the Denisovans of Asia, was the ability to communicate more efficiently and systematically.[19] These advanced communication skills—which went beyond mere grunts and pointing—enabled *Homo sapiens* to prepare for confrontations with violent tribes, climate change, even famine. Later on, vocal language was translated into written language, which allowed humans to free up mental energies to develop more complex technologies. Think about it: I don't have to remember *all* the information; I just have to remember *where* the information is stored so I can access it. These communication skills allowed humans to build off of the knowledge of past discoveries, enabling *Homo sapiens* the ability to rapidly adapt to changing environments.

We humans are a resilient and special bunch. We've been around for two hundred thousand years, and around seventy thousand years of those have been similar to what we know today as "behaviorally modern human."[20] One of the biggest events that showcases our resiliency happened around 75 thousand years ago at Mount Toba near present-day Indonesia. It is thought that an eruption occurred that covered the sky with ash, killing numerous plants and animals, reducing the starving, struggling humans to a mere few thousand in number. According to genetic studies, our population possibly got down to as low as ten thousand in Africa.[21] These are the numbers of near-extinction. Yet, as we've seen with previous catastrophes, this low point in human history was likely the catalyst that was needed to refine

---

[19] We can think of the Denisovans as the European Neanderthals' counterpart from Asia.
[20] Henry M. McHenry, "Human Evolution," in Ruse and Travis, *Evolution*, 265.
[21] Darrel Falk, "Human Origins: A Summary of Scientific Backdrop for Theological Discussion," lecture at BioLogos: Re-Imagining the Intersection of Evolution and the Fall, Chicago, March 28, 2015.

not only how humans behaved but also who *Homo sapiens* were.[22] And so, steadily growing in number, groups of humans spread out of Africa around sixty thousand years ago and started to colonize the globe in ways that we know today. Thus, it is important to remember that we are truly the most unique species of organism to ever exist. In the short time *Homo sapiens* have been on earth, we have advanced at a rapid pace with no end in sight. We are truly special indeed.

---

[22] As mentioned previously, we might look to Stephen Jay Gould's concept of punctuated equilibrium. See Stephen Jay Gould, *Punctuated Equilibrium* (Cambridge, MA: Belknap Press, 2007), 54.

5

RELATING TO SCIENCE

SOME THINGS—NOT ALL THINGS—NATURALLY integrate into one smooth system. A few years ago we were potty training our two-year-old daughter at the time when a curious smell began to waft from the room she shared with her four-year-old brother, who slept on the top bunk, with her just below. For several days, every time we walked by their bedroom we would smell the odious aroma of the dreaded wet bed. "Honey," we inquired, "did you have an accident in your bed?" She assured us that she hadn't—and her dry bed supported her claim. It was driving her mother and me mad—we knew the smell was coming from their room, but for the life of us we couldn't figure out where it was originating from, even after tearing apart their bedroom. One evening after my wife went to work, I was getting the kids ready for bed when I heard my son running his fingers along the air register on the floor. I stood in the doorway, watching him crouch above the register, repeatedly running his fingers over it, never breaking his attention from it. He must have felt me staring at him because out of the blue he murmured in a soft, shameful voice, "I didn't pee in the vent."

"Excuse me? What did you say?" I was in a state of bewilderment.

Now looking up at me, waiting to see if I took the bait, he said, "I *didn't* pee in the vent."

"Buddy," I replied, "I didn't *ask you* if you peed in the vent, but I will now. Buddy, be honest with me, did you pee in the vent?!"

"Yes, but it's okay, Dad," he said with a touch of childlike belief, "the vent and toilet go to the same place."

As I began wondering how I'd possibly clean up such a mess, I also began to wish that what my son had said were true—that the plumbing and the ductwork were indeed integrated systems. But not all systems can be integrated; plumbing and ductwork are two such systems, and for good reason. Unfortunately, it seems we often treat science and faith like plumbing and ductwork—systems that shouldn't (or can't) be integrated with one another. However, I think faith and science, despite the fact that they occasionally seem at odds, can be integrated into a cohesive system without having to be in conflict with one another.

In the 1960s Ian Barbour developed a typology for categorizing how faith and science typically interact with one another.[1] In *Religion and Science* Barbour contends that there are several ways this interaction can happen; two of these are helpful for our purposes in this book: *conflict* and *integration*. Because integration is the category that I feel is the most helpful to the Christian faith, while simultaneously being truthful to our Christian narrative, I will first quickly comment on the history of conflict between faith and science.

There has been a long history of struggle between science and faith—particularly within the Christian religion. This struggle should come as no surprise given that, for most of Christianity's history, the church was not only a spiritual power but also a temporal power. If I had to point to why the conflict between faith and science persists even today, I would probably say that it is because of a power struggle. It's like raising children: when they fight, it's rarely about the toy or possession they're bickering over. It's almost always about power. "He has my doll!" my daughter would scream. "But you weren't playing with your doll," I would say with complete peace and serenity. "But it's *mine* and *he* can't have it!" Ah, there's the crux of the issue. It seems to me that in much the same way the Christian church and scientists have had one conflict after another.

---

[1]See Ian G. Barbour, *Issues in Science and Religion* (Englewood Cliffs, NJ: Prentice-Hall, 1966).

Galileo's story is one of the most prominent examples of scientific and religious conflict. Labeled the "father of modern science," Galileo was born in Italy in 1564, the Protestant Reformation still ablaze. It was a tense time for Christianity in Europe. Even though he was a devout Catholic, Galileo started to question older Catholic assumptions about the natural world. He was advocating Copernicus's heliocentric model of the universe (in which the earth revolves around the sun), while the Catholic Church sided with others on a geocentric model (in which the sun revolves around the earth).

We might wonder why the Catholic Church even cared about this debate. While the answer to that question is complex, a major reason a heliocentric universe was alarming to the Catholic Church was that it threatened to take humans out of the center of the universe—everything might not, quite literally, revolve around us.[2] After Galileo was forced to stand trial under inquisition for promoting these heliocentric beliefs—and despite years of observation and evidence supporting his claims—the church handed down a judgment concerning Galileo's idea that the sun was "motionless" (i.e., not revolving around the earth). On February 24, 1616, the Catholic Church declared,

> All Father Theologians said that this proposition is foolish and absurd in philosophy, and formally heretical since it explicitly contradicts in many places the sense of Holy Scripture, according to the literal meaning of the words and according to the common interpretation and understanding of the Holy Fathers and the doctors of theology.[3]

In other words, if Galileo didn't recant his heliocentric belief, he'd be in some serious trouble.

But there is much debate, even today, over whether or not this conflict was between faith and science or actually about sovereignty. To

---

[2] It should be noted that the church was particularly concerned about Galileo—a layman—telling the clergy how to interpret the Bible.
[3] Maurice A. Finocchiaro, *The Galileo Affair: A Documentary History*, California Studies in the History of Science (Berkeley: University of California, 1989), 146.

put it differently, it may in fact have been a power struggle, a conflict over who got to decide the veracity of scientific claims—the scientific community or the Church. A few years after the Church condemned Galileo as a heretic, he published papers against Aristotle's perfect universe system. Galileo's *Dialogue Concerning the Two Chief World Systems* was a direct stab at the power structure of the day. The Catholic Church at the time (and still today) relied heavily on Thomas Aquinas's theology, which, to be overly simplistic, combined Aristotelean philosophy with Augustinian theology. So any attack against the Aristotelean universe was seen as a direct attack against the Church. What Galileo was doing was dangerous, and it was met with great suspicion. Nevertheless, despite the Catholic Church's suspicion, it's important to pause here and make the distinction that it was in fact Galileo's faith in a Creator who perfectly ordered the universe that led him to pursue his research. He wanted to find meaning, beauty, and order in God's creation. So, though the Church may have failed him, God did not.

A similar sort of struggle over the "sovereignty of truth" is happening with the hotly contested debates between creationists and evolutionists—or more precisely young earth creationists and ultra-Darwinists. Broadly speaking, young earth creationists are those who believe that the earth is between six thousand and ten thousand years old and that life did not evolve by natural selection. In general, ultra-Darwinists are those who often view evolution to be the final nail in God's coffin. One of the most helpful books I've read on the struggle between young earth creationists and ultra-Darwinists is *Darwin's Pious Idea* by Conor Cunningham.[4] In it, he says that there is no actual conflict between Christianity and evolution; instead, the conflict is really between young earth creationists and ultra-Darwinists. Both of these groups tend to highjack the conversation for their own power plays, much like the seventeenth-century Catholic Church did with Galileo.

---

[4] Conor Cunningham, *Darwin's Pious Idea: Why the Ultra-Darwinists and Creationists Both Get It Wrong* (Grand Rapids: Eerdmans, 2010).

Both have strong ideologies that stifle communication, both lack the listening skills to hear each other and understand opposing arguments. Young earth creationists often have an unwavering literal interpretation of certain biblical passages, especially the creation narrative found in Genesis 1. Ultra-Darwinists often have a hidden agenda to create atheists (or not-so-hidden in the case of spokesmen such as Richard Dawkins). And both sides talk out of turn.

HAVING ALL THE ANSWERS

Like most firstborns, when my eldest child was a little boy, he knew everything. I mean everything. I was talking with a friend while our boys were watching *Star Wars*, and we overheard my son explaining to his buddy who Darth Vader was. "They're sister and brother," I heard him wisely explain to his friend. "Darth Vader is Princess Leia's sibling." Ah, not so much, buddy. See, my son had *some* of the information—but not all—and decided that he knew the whole story. Instead of fact checking and listening to others about Darth Vader's lineage, he just grabbed half of the story and filled in the gaps. If I had tried to correct him, he would have been unwilling to acknowledge his error. As we've all experienced, it's hard to teach a person what they believe they already know.

The same kind of thing often happens with young earth creationists and ultra-Darwinists—the old "I'm right; what's the question?" attitude—and it is what ultimately fuels the conflict between them. We hear major voices in the debate doing the exact same thing as a seven-year-old—we only need to look to Ken Ham on the young earth creationist side and Richard Dawkins on the ultra-Darwinist side to find fitting examples. Ham typically claims that most of the science about an old earth and macroevolution is incorrect—or even worse—deceptive—insisting instead that the biblical flood, for instance, can explain why the earth seems extremely old. But there's a problem. While a team of scientists surrounds him, many lack convincing scientific credentials.[5] The rest of the scientific community incomparably dwarfs the scientific credentials of young earth creationists. In other words, Ham's group lacks solid authority to talk about science from a scientific perspective.

Would you trust your cousin Mike who teaches high school English to diagnose your ailing minivan—his only relevant credentials being that he also owns a minivan? Or would you trust the certified mechanic at that minivan's dealership? Who would you hire to repair it? Mike or the mechanic? See, we rarely quibble over specialized knowledge in practical matters. But when it comes to what we perceive to be issues of faith or morality, we're often content siding with the perspective that makes us most comfortable. If Mike tells me my engine is fine, I'm probably going to be comfortable because now I'm not staring down a $2,500 repair. But when I find myself on the side of the road because my minivan's timing belt has snapped and destroyed my engine, I'll be wishing I'd listened to the mechanic.

To be clear, young earth creationists are not alone in this behavior. Ultra-Darwinists likewise often lack credentials to speak about metaphysics, philosophy, and theology. Richard Dawkins is a good example of a person who assumes the mantle of expertise on such issues, yet his sloppiness becomes evident as he effortlessly and frequently slips into philosophy and suprascientific ideas in most of his writings. Dawkins, trained in evolutionary biology, has a lot to say about religion (and that's an incredible understatement!). He's penned books about how God doesn't exist, given numerous talks about how evolution proves that there is no God, and frequently comments on how the Bible is a how-to book for committing moral atrocities. In the same way that Ham isn't qualified to discuss the veracity of scientific

---

[5] I need to acknowledge that there are some scientists with good credentials on that side of the debate. One area where they repeatedly make a strong point is about how starting points and presuppositions affect scientific research. Still, many of the scientists that hold a young earth perspective are not adequately peer reviewed—a process that tries to hold scientists accountable.

discoveries, Dawkins has not earned the right to be heard when it comes to metaphysics, philosophy, and theology.

I hope the reader understands that I'm not attempting to ruffle feathers or be overly critical in my assessment of ultra-Darwinists and young earth creationists. I wish only to point out that, while these extreme positions are surely in conflict with one another, they are neither representative of most people who find the theory of evolution credible nor representative of most Christians around the world. The ultra-Darwinists and young earth creationists are loud, vocal, and often hold aggressive positions, but they are assuredly in the minority.

HEALTHY INTEGRATION: A LITTLE HELP FROM HEGEL

It might sound strange to any fans of philosophy out there, but I think one of the most illustrative examples of a healthy way to approach the integration of science and faith can come from the philosopher Georg Wilhelm Friedrich Hegel. If you've never read any of Hegel's original writings, some would be inclined to consider you fortunate! While much of the philosophy of today owes a significant debt to this great thinker, he's notorious for being obtuse and is laborious to read. Nevertheless, he has some important illustrations for faith and science integration.

After building off of previous philosophers, Hegel formed what is now known as the Hegelian dialectical method, which goes something like this: first, you have *thesis* (or the original idea), then an *antithesis* (a conflicting idea), and finally a *synthesis* (some kind of integration of the previous two). Most people fight over the interpretation of Hegel's concept of synthesis. There are two prevalent interpretations of the Hegelian synthesis that are pertinent for resolving the conflict between faith and science.

First, in what I call "the blended model," the two opposing viewpoints are blurred together, forming a new position. In our case, faith and science become a kind of metaphysical-scientific Frankenstein and get stuck in a perennial cycle of thesis and antithesis—each weary and fearful of the other, tightening grips, and struggling to remain in control and retain power. Synthesis looks to move the thesis and antithesis to common ground but does so by blending the positions together to the point of indistinctness. A synthesis of something—in our case, faith and science—can often blur the two entities together in much the same way as when someone colors over a blue crayon with a red one. The original thesis and antithesis are lost, and a new color emerges. But Hegel's word *synthesis*, which is anemic at best for what we're trying to do with integrating faith and science, can be made clearer by looking to Hegel's original language.

When we translate from one language to another, we often lose nuanced meaning. For instance, it is quite common to find the German word *Aufheben* translated into English as *synthesis*, a central word used to describe his dialectic. However, *Aufheben* may also be translated as "sublation." Understanding the difference between synthesis and sublation is helpful in understanding the differences between healthy and unhealthy integration of faith and science. Instead of a blending that leads to indistinctness, where the original positions of faith and science are lost, sublation—which I believe is a truer translation of Hegel's *Aufheben*—means that the thesis and antithesis are in constant tension with each other. *Synthesis* seems inadequate because it places emphasis on the elimination of the original distinctness of the thesis and antithesis, when in truth Hegel seems to be arguing not that the synthesis is the elimination of distinct parts of thesis and antithesis, but the assimilation of these distinct parts into a coherent, sublated whole; a whole that has not forgotten its origin in the constant tension of the thesis and antithesis.

Recasting Hegel's dialectic in this way allows us to think of the sublation—or what I want to call *integration*—of faith and science as doing three things: (1) It preserves the unique identities of the thesis and the antithesis. In our case, a healthy integration or sublation can allow science to remain science and faith to remain religion. In other words, they do not need to blend into an indistinctness that negates the value of either science or faith. (2) In a paradoxical way, sublation also changes each identity. I've heard of some wedding traditions where the bride and groom both bring a little bag of sand and drop a pinch into each other's bag. Both bags become different yet retain their original identities. If faith and science experience integration in a healthy way, it will be impossible for each to remain unchanged by the other. (3) Sublation can, and does, advance each identity—one mutually benefiting the other. My wife and I had traditional unity candles at our wedding, the kind where each of our parents lit a candle representing the beginning of our lives. During the service we simultaneously lit a third candle, representing our new life together—two distinct families coming together as one in our marriage. The unity candle tradition typically requires the couple to blow out the original candles that were lit by the parents—emphasizing that the old lives have been replaced by the new. Instead, we opted to leave all three candles lit. We weren't leaving behind the influence of our families—forgetting where we came from and moving on to a new life. Not at all! Rather, we appreciated our families, our histories, our distinct identities. But that also didn't mean we weren't excited about the new life we would shape together. And so, with the three candles lit, the stage was illumined with more light than there previously had been in the service—a powerful symbol of the unity of our families through our marriage.

For healthy integration to be compatible, we must understand faith and science as always advancing each original identity while dwelling within mutual submission to one another as areas of expertise are acknowledged. Thus, in the case of faith and science, when we integrate them through sublation, the light becomes stronger because, frankly, it moves closer to the truth, and the truth is greater than any one human paradigm.

Another great example that illustrates what the healthy integration of faith and science looks like is that of the two natures of Christ—a case Hegel himself employs when talking about sublation in his *Lectures on the Philosophy of Religion*.[6] This important orthodox doctrine preserves the two distinct natures of Christ while at the same time recognizing the paradox of his natures. One can juxtapose this doctrine against the heresy of monophysitism, which claims that Christ has only one nature. Orthodox views of Christ's are careful to articulate that his two natures are unmixed—he is both fully God and fully human. And in much the same way we need to independently preserve faith and science while at the same time acknowledging their integrated relationship. If we follow Hegel's example of healthy integration, it will go a long way toward helping us understand how to preserve the integrity of both science and faith.

In addition to understanding integration by way of sublation, Hans Küng articulates another helpful way of looking at it in his "mutual critical correlation and confrontation" method, which sounds more intimidating than it is. In his book *Theology for the Third Millennium*, Küng drives home the idea that in order to remain authentic to the original ideas (e.g., faith and science), it is imperative for there to be critical debate and not just correlation.[7] When science and faith interact with each other, there should be no room for dodging questions, for talking past each other, for ignoring the truth—come what may. Real learning takes place with free dialogue and *mutual critical* exchange. That means that we shouldn't just expect theology to ask critical questions of science but science to ask critical questions of theology.

French philosopher Simone Weil should assuage some of our fears about true critical engagement with science when she says,

> It seemed to me certain, and I still think so today, that one can never wrestle enough with God if one does so out of pure regard

---

[6]Georg Wilhelm Friedrich Hegel, *Lectures on the Philosophy of Religion*, ed. Peter C. Hodgson, trans. P. C. Hodgson R. F. Brown, J. M. Stewart, one-volume ed. (Berkeley: University of California Press, 1988).
[7]Hans Küng, *Theology for the Third Millennium: An Ecumenical View* (New York: Doubleday, 1988).

for the truth. Christ likes us to prefer truth to him because, before being Christ, he is truth. If one turns aside from him to go toward the truth, one will not go far before falling into his arms.[8]

If we desire science and faith to experience genuine integration and critical engagement, then we ought not fear the discoveries of scientists that may seem to contradict tenets of our faith. Likewise, we ought not fear the careful descriptions of theologians because they may seem to contradict the world we see before us. We should never fear moving closer to the truth.

BUT WHERE ARE WE GROUNDING OURSELVES?

Establishing our beliefs on firm ground is extremely important. I remember when one of my sons thought it would be a brilliant idea to jump off of a bed onto a large exercise ball. He thought he would be able to bounce off of it just like they do in the cartoons. He bounced off of the ball, alright; he just didn't land on his feet. He replaced firm ground with something much less stable and suffered dearly for it. If we are going to attempt to reside in Barbour's mode of integration between science and faith, then it is imperative that we operate from within a tradition that can help hold both science and faith accountable. Thus, I believe constructing a solid theological foundation to stand on is essential.

Every Christian is grounded in some kind of tradition despite our tendency not to recognize such at times. Whether we are nondenominational, Baptist, Pentecostal, or Roman Catholic, our faith has been shaped by a certain perspective, one often grounded in a robust tradition. Our first major step in understanding science has to be to understand our own history. What do the theologians in my tradition say about its own doctrines? What do they say about scientific advancement and discovery? More importantly, what are the elements

---

[8]Simone Weil, *Waiting for God* (New York: Putnam, 1951), 27.

within my own faith that are primary and what are secondary? In other words, what are the nonnegotiables of your Christian tradition, and on what things does your tradition reserve its opinion; that is, on what things is it willing to be flexible?

For instance, most Christians who fall under "orthodox" Christianity —this would include most Protestant denominations, the Roman Catholic faith, and the Eastern Orthodox Church—have a few fixed doctrines that mainly revolve around who Jesus is: Trinity, virgin birth, incarnation, dual nature of Christ, bodily resurrection (of Jesus and us), and the like. Other beliefs in the faith—for example, one's position on end times and creation—often have broader, less formulated doctrines. Most denominations, my own Free Methodist Church included, take a position that goes something like this, but with a little more theological flair: although we don't exactly know how, we believe that God created the universe, and God will make all things new in the end—however he chooses. This openness provides believers standing from within this tradition a bit more flexibility in interpreting Scripture than those dominations with more elaborately worked-out and prescriptive doctrines.

If you happen to be in a church or denomination whose doctrine is more prescriptive about scientific findings, I would encourage you to keep asking hard questions. All doctrine should be able to be tested by difficult questions. But remember that asking difficult questions often means we'll have to work through difficult answers. And, in the words of Wendell Berry, sometimes we need to "ask the questions that have no answers."[9] Asking these kinds of questions often leads us to deep reflection that spans many years.

In fact, the Christian faith has a long history of asking difficult questions. Take, for example, one of the early Christian fathers, Augustine of Hippo. He was remarkably outspoken about how to interpret passages

---

[9]This is in the poem "Manifesto: The Mad Farmer Liberation Front." See Wendell Berry, *The Country of Marriage* (Berkeley, CA: Counterpoint, 2013), 14-15.

of Scripture in light of new questions and evidence. Augustine acknowledged that the first few pages of Genesis were particularly challenging to interpret and, for the sake of evangelism, should not be interpreted literally *ipso facto*.[10] We've also already looked at Galileo's controversy with the Roman Catholic Church. During his house arrest he wrote an important and famous letter to the Grand Duchess Christina of Florence. In 1613 Duchess Christina asked an honest, hard question about the biblical accuracy of the heliocentric model. What followed was a fabulous discussion concerning God's revelation, among other scientific topics. In these letters Galileo articulates the "two books" notion of revelation—Scripture and nature—that became the model for how people of faith could use nature and science to better understand God. Here, Galileo says to the Duchess, "I do not feel obliged to believe that the same God who has endowed us with senses, reason and intellect has intended us to forego their use and by some other means to give us knowledge which we can attain by them."[11] In our own contemporary time, we can learn from these past individuals and see science as a vehicle by which to better grasp our own faith and relationship with God, thus starting a healthy integrative discussion.

SEEKING TRUTH

Finally, we must remember that what true science and faith have in common—especially when they are unadulterated and without hidden agendas or power plays—is the desire to seek truth. Though both come to truths (some alike, some different), often via unique methods, they are both ultimately concerned with leading us toward the truest understanding we might have of our world and reality.

But in our age of scientism and technology, we have to be careful with our relationship to science. Even when practicing healthy integration, we should not come to a place where we are unable to defy

---

[10]See Augustine, *The Literal Meaning of Genesis*.
[11]Galileo Galilei, "Letter to the Grand Duchess Christina of Tuscany," 1615.

science, where it cannot be questioned. The most recent century in our world serves as an important warning against the worship of science—we have been blessed with the progress of science in some ways and cursed by its progress in others. We need only look at the development of nuclear technologies for our warning. For the first time in the universe's history, humans willfully developed a technology with the clear knowledge that it had the power to exterminate all life on our world. While many prominent figures in the twentieth century questioned the ethics of nuclear bombs, those voices lost influence in the face of those who claimed they were necessary. We will only be able to practice healthy integration insofar as we are willing to be courageous, holding both faith and science accountable to one another and to the communities that so depend on them.

PART THREE

AN INTEGRATED APPROACH TO EVOLUTION AND THE CHRISTIAN FAITH

6

UNDERSTANDING EVOLUTIONARY THEORY CAN BE EMPOWERING

I WAS STANDING IN THE CHECKOUT LINE at the grocery story with one of my sons, who was young at the time. As I was unloading the cart onto the belt, I looked over at my son, who was staring at a magazine cover. I certainly don't expect much in the way of morality from tabloids, but the one that my son was looking at had a particularly revealing image of a woman on the front. Trying to play it cool and not make the incident worse than it had to be, I patted him on the shoulder and said, "Look at this toy over here. Pretty neat, huh?" My son didn't move. In fact, like a deer in the headlights, he was motionless in front of the magazine rack.

"Is she a mommy?" he asked, looking at the busty figure.

Oh boy, how do I respond to that? I patted him on the back again and said a bit more forcefully, "Come on bud, let's not look at that." But he didn't budge—totally frozen to the floor. This time I grabbed his shoulder and started pulling him away and tried to explain how we shouldn't look at that kind of thing. And as I was dragging him from the tabloid, he just stared at the magazine and said, not in a voice of defiance but of pitiful sadness, "But I *want* to look!"

While I was struck with how honest and open my child was being, I couldn't help but wonder what was going on under the surface. Where did that come from? Why was my son—who normally couldn't sit still if his life had depended on it—standing there in a catatonic state of visual consumption? What I now know is that he was encountering something that all adult men intuitively come to experience; my son was seeing something that was visually stimulating coupled with a sensation of testosterone. This surge of hormone—which men experience at a rate tenfold that of women throughout their lives—pulled him toward the tabloid stand. And much like a game of tug of war at a family reunion, he was caught between a dad who wanted him to look away and a body that wanted to stay right where it was.

He was far too young, but if he only knew the destructive power that those images can bring and where his urges came from, then he could more easily reorient his actions toward healthier—and holier—living. Now, is simply having knowledge about where we come from a perfect shield from sin? Of course not. But we first have to know where we come from before we can get to where we want to go.

KNOWLEDGE IS POWER TO CHANGE

An essential benefit of adopting an integrative understanding of evolutionary theory and our Christian faith is that we become better able to recognize who we are and why we tend to have certain behaviors. And this knowledge of our biological and environmental pasts may actually lead us to become more whole—and holier—individuals.

Not too long ago I was at the doctor for a regular checkup, and he noticed a weird blemish on my skin. He started to ask mild questions that became more and more serious as the conversation went on: "Have you noticed this spot on your back changing?"

"Yeah, I guess so."

"How long has the spot been there?"

"Uh, I don't know." Then I noticed the questions began to change.

"Is there a history of cancer in your family?"

"Well, I guess so. My brother had melanoma on his leg."

"Matt," my doctor said, "why didn't you tell me that from the beginning?!"

What I wasn't aware of at the time, but I am painfully aware of now, is that my family's medical history—not just mine but my relatives' and ancestors'—really mattered for assessing current problems in my health. Thankfully everything worked out fine, but this interaction with my doctor was a tangible example to me of how important my past is. When we have knowledge of our roots, we have knowledge of our current situation. When thinking of how our ancestry affects our lives today, evolutionary biologist David Sloan Wilson observes,

> Our unique attributes evolved over a period of roughly 6 million years. They represent modifications of great ape attributes that are roughly 10 million years old, primate attributes that are roughly 55 million years old, mammalian attributes that are roughly 245 million years old, vertebrate attributes that are roughly 600 million years old, and attributes of nucleated cells that are perhaps 1,500 million years old. If you think it is unnecessary to go that far back in the tree of life to understand our own attributes, consider the humbling fact that we share with nematodes (tiny wormlike creatures) the same gene that controls appetite. At most, our unique attributes are like an addition onto a vast multiroom mansion. It is sheer hubris to think that we can ignore all but the newest room.[1]

If we indeed come from a long history of ancestors—some of whom include animals that, right now, seem very distant to us—then just as a medical doctor might ask us if cancer runs in our family, we might ask questions about behavioral traits that run in our lineage. In a

---

[1] David Sloan Wilson, *Evolution for Everyone: How Darwin's Theory Can Change the Way We Think About Our Lives* (New York: Delta Press, 2007), 70.

strange but understandable way, it actually matters what the nematode appetite was like. It matters that our hunter-gatherer male relatives had frequent sex and abandoned families in search of more mates. And it matters that our ancient female relatives decided to feed their offspring at the expense of other hungry children. Thus, a major step in developing a vibrant Christian life is recognizing that our biological and environmental proclivities that express themselves in negative and positive ways are steeped in our ancestral history.

Michael Dowd, the author of *Thank God for Evolution*, has a helpful way of presenting why understanding our roots and teaching it to our children is of the utmost importance for Christian development. In the book Dowd discusses his idea of *mismatched instincts*—the concept where one's natural tendencies don't align with modern life—coupled with what evolutionary psychologist Deirdre Barrett calls *supernormal stimuli*—the notion of unnatural and exaggerated temptations.[2] Together, this is what Dowd and Barrett are saying: many of our instincts, from the desire to eat to the drive to have sex, were incredibly important to our ancestors—both human and prehuman. It's because of those instincts that we are alive today. We should be thankful for them! Sometimes I remind my classes, most of which are made up by nineteen- to twenty-one-year-olds, that they should be thankful that their grandparents got frisky forty years before they were born. When I make that pronouncement, their reaction is unanimous: "Ah, Dr. Hill, we don't want to think about that!" If I'm being honest, half of the reason I say it is to get that reaction. The other half is that I'm very serious. If our grandparents didn't have strong libidos, we wouldn't be alive today to be disgusted at the thought of their sex drive. Those drives and desires, from base urges to complex inclinations, are what

---

[2] For this section I am very grateful for Michael Dowd's work. Most of the ideas in this section come back to him one way or another. For more information I encourage the reader to see Michael Dowd, *Thank God for Evolution! How the Marriage of Science and Religion Will Transform Your Life and Our World* (Tulsa, OK: Council Oak Books, 2007), 155. See also Deirdre Barrett, *Supernormal Stimuli: How Primal Urges Overran Their Evolutionary Purpose* (New York: W. W. Norton, 2010).

fueled human development and allowed us to out compete other organisms who didn't possess the same instincts. In other words, the ancestors of those who didn't want to eat all the time and have sex all the time are not here to tell their stories. Humans, by cultivating some proclivities from natural selection, can develop a practice of restraint. Here I use food and sex as main examples because they are, quite obviously, two drives that are common and prominent among us. Indeed, there are a host of other instinctual drives we could also discuss: feelings of empathy, acts of altruism or self-preservation, and violence.

While it's obvious that we have these instincts, sometimes they seem out of place given contemporary circumstances. For instance, child mortality rates have plummeted in just the last 150 years, making unnecessary, from a reproductive standpoint, the need to conceive as often as possible. Yet this doesn't mean that the incessant drive to *try* to conceive a child has gone away. It's like a retired race-car driver who—still recalling the days when he needed a powerful engine for racing—drives around his block in his Bugatti. The *desire* to race is still part of his identity, but the *need* to race has long passed.

While not all our drives are antiquated like this—as some evolutionary drives are positive in nature, even today—the same logic applies to the consumption of food today. While famines and starvation are very real for many people in two-thirds of the world—a concern we ought not overlook—most of the problem with contemporary hunger has to do with food distribution, not food shortages. Humankind is producing an amount of food that far surpasses what history has seen before. A few hundred years ago, people had to be able to retain fat so that when winter or droughts came they wouldn't become emaciated or starve. Today we have immense storage structures and methods to protect against this threat. Before modern advances in agriculture, our food storage was put away in our stomachs, waists, and thighs. In other words, being able to get "fat for the winter" was a real thing. Now we just get fat.

It cannot be overstated that the instincts of *Homo sapiens* who were living ages ago were incredibly necessary for helping us survive to reproductive age. But now, in the twenty-first century, many of these instincts are mismatched given advances in human technology, agriculture, and medicine. We simply don't *need* to eat all the time or have sex all the time. As a result of such changes, there is now a need for temperate behavior—behavior that, in an ironic way, does not come naturally to us.[3]

FRIES ARE THE PORNOGRAPHY OF FOOD

Many of our mismatched instincts, when combined with supernormal stimuli, are a recipe for moral and spiritual disaster—that is, for unholiness. If you're wondering what Barrett means by "supernormal stimuli," let me explain it with an example.

Burger King french fries are the pornography of food. Here's what I mean: Have you ever walked by a room where someone was cooking bacon? It's absolutely irresistible. You start salivating and looking around to figure out where the smell is coming from. There are really good reasons for this. Bacon is high in protein and fat and has the calories to provide energy for a long time. As we just discussed, our ancestors were selected after generations had lived and died for such sustenance. So, in a sense, we are hardwired to love fatty foods. We're *almost*—and this word is important—a *slave* to fatty foods. And Burger King french fries are our food master. I remember when my friend and I were sixteen and Burger King had just put out a new style of french fries. They figured out something about evolution and capitalized on

---

[3] I do not intend to offend those who are more studied in the area of biology compared to the average reader. I am well aware, as my previous book can attest, that we have numerous prosocial and altruistic tendencies from our evolutionary roots that enable us to live harmoniously in groups and act kindly and even good. For the sake of the flow and narrative of this book—not to mention the audience I am writing to—I want to establish some of the basic ideas of evolution before I discuss the rule-breakers and caveats. This approach may lead to some oversimplifications at points. It is my argument in this chapter, and indeed the argument in my previous work, *Evolution and Holiness*, to display how Christians can nurture such ancient sociobiological proclivities to live a holier life.

it. Instead of having more potato as their new ingredient, they had more oil and fat—a nearly irresistible temptation to our sixteen-year-old instincts. They capitalized on the fact that we were hardwired to search after fat, salt, and sugar.

I say that french fries are the pornography of food because both fries and pornography work on similar human desires. French fries are foodlike substances that are ultimately unhealthy for us; pornography is a sexlike icon that is likewise unhealthy for us. French fries have all the glamorous and seductive parts of food—the fat, the salt, the sugar—but none of the substance that provides long-lasting enjoyment or health. Pornography, in much the same way, possesses the glamorous and seductive aspects of a sexual relationship, but is at its core shallow and vapid—it is neither sex nor a relationship. It shouldn't surprise us then that both french fries and pornography are highly addictive. As a pastor and a professor working with college students, I can't tell you how many times I have had to counsel people who have dependencies on food or pornography. Whether it be health complications or broken families, the supernormal stimuli that contemporary humankind faces are very destructive.

Remember, these supernormal stimuli are just that: supernormal. They aren't natural. They're conjured up in a lab full of people who know how to manipulate in us these ancient but no longer necessary proclivities. Advertising agencies know that *Homo sapiens* have evolved to desire copious amounts of sex and be attracted to visual stimuli. So sex is used to sell everything from blue jeans to bicycles. The fast food industry knows that *Homo sapiens* have needed fat, salt, and sugar to survive—that we fall into a kind of trance when we smell bacon or fries—and so they sell food products that are nearly irresistible. Humans are in a predicament. We exist in a world in which we encounter both our mismatched instincts from ages ago and the supernormal stimuli of the twenty-first century. Barrett expresses the problem in this way:

The most dangerous aspect of our modern diet arises from our ability to refine food. This is the link to drug, alcohol, and tobacco addictions. Coca doesn't give South American Indians health problems when they brew or chew it. No one's ruined their life eating poppy seeds. When grapes and grains were fermented lightly and occasionally, they presented a healthy pleasure, not a hazard. Salt, fat, sugar, and starch are not harmful in their natural contexts. It's our modern ability to concentrate things like cocaine, heroin, alcohol—and food components—that turns them into a menace that our body is hardwired to crave.[4]

See, our problem is a chimera of sorts—a Greek mythical multifaceted creature composed of mismatched instincts combine with supernormal stimuli. It is a cocktail of temptation unprecedented in humankind's history. This chimera is difficult to combat, and—besides contributing to our gluttony, lust, self-indulgence, and other negative proclivities—it leads us to develop habits in our lives that pull us away from Christian virtues, from holy living. But combat it we must, and the more we learn about where we come from, the more we are able to do so. We believe God's Spirit dwells with us, that we can be transformed and renewed; we can become better than we are. We are not slaves to our biology, no matter how strong negative instincts seem to be.

THE LIFE OF BRAIN(S)

One of the most important steps in learning about our biological lineage is taking a look at the evolution of *Homo sapiens* brains. While it's imperative for us to note that people are more than just their brains, our personalities, moral decision-making, and the ability to have self-control are cerebral activities. In other words, while we certainly are whole persons who cannot be reduced to mere brain activity, the brain is essential to human viability—we can live without

---

[4]Deirdre Barrett, *Waistland: A (R)Evolutionary View of Our Weight and Fitness Crisis* (New York: W. W. Norton, 2007), 26.

an arm or a leg but not without the gray matter within our skulls. For very good reasons this organ is paramount to our personhood. Understanding the various parts of the brain can give us a window into our drives and behaviors.

If we're able to recognize and respect where these instincts come from, we can in time learn to control them. Sometimes I get *hangry*— a potent combination of hunger and anger. My wife teases me a bit when, usually right before dinnertime, I pace around the house like a caged animal. She'll call out to me in a patronizing voice (which I completely deserve), "Matt, are you feeling a little hangry?" She's right, as usual, and it reminds me to take a step back and either muster up some self-control or grab a bite to eat. This acknowledgment of where my frustration is coming from is helpful in navigating my way toward becoming a better—that is, more disciplined and holier—person.

Homo sapiens' brains have evolved over millions of years in a curious way.[5] As time went on, newer and more complex sections of the brain were added through the process of evolution. In the same way a geologist might look at cross sections of the earth's crust to determine the age of the fossils found in segments of the soil, we can look at the sections of the brain and have a very good understanding of which ancestral root each part of the brain has grown from. From an evolutionary perspective the parts of our brain range from very old to relatively new. Each section of the brain has certain functions it performs in relation to the other. Some segments override other portions of the

---

[5]After scouring ways to figure out how to best describe our brain function, I really can't find any more helpful examples and explanations than from wife and husband Connie Barlow and Michael Dowd. I'll explain the gist of what they say in this section, but if you're interested in further readings and presentations from them, please see these beneficial resources. After having many conversations with him I've found that besides being an incredibly hospitable and generous guy, Michael Dowd's illustrations and explanations are helpful because they are simple, thoughtful, and incredibly illustrative in communicating these ideas to the audience that this book wishes to reach. Please note that most of this section should be sourced back to either *Thank God for Evolution!* or various presentations by Dowd and Barlow on evolutionary psychology. See Dowd, *Thank God for Evolution!* The following is one presentation on a similar topic: Michael Dowd, "Why We Struggle Now," TEDx, June 14, 2012, www.youtube.com/watch?v=DDMOF7qtlh8.

brain for the benefit of the person; some parts encourage other parts to secrete more or fewer hormones; some parts take inventory of social situations and encourage prosocial or antisocial behavior. But all parts of the brain are moment by moment negotiating with each other over the kinds of decisions we make.[6]

The most ancient part of the brain—sometimes called the *reptilian* part—is where our basest urges reside. This part of the brain is composed of the brain stem and cerebellum, wherein lie the rudimentary instincts of survival and reproduction. Some say the reptilian part of the brain is concerned only with the three S's: safety, sustenance, and sex; it is the part of the brain that's kept us alive and evolving for countless generations. The base instincts that come from the brainstem were important in our evolutionary history and still are today. Those who suppress the reptilian part of the brain perish very quickly because we need to be safe in order to find food and shelter, need to eat to live, and need to have sex to reproduce. Yet, as I'll discuss in a moment, when the reptilian part of the brain is out of balance with the other portions, we tend to act impulsively and with little self-control.

The *old-mammalian* part of our brain—or what scientists call the hippocampus, the amygdala, and the hypothalamus—has been a part of mammals' brains since their inception. Whereas the reptilian part of the brain was premammalian, connecting us with much of the organic world, the old-mammalian part is shared nearly exclusively with mammals, which tend to be social creatures. Instead of caring about finding a mate or sniffing out food, the old-mammalian part of our brain is concerned with empathy, social connections, reciprocal relationships, bonding, and emotions. Of course, more organisms than just mammals are social—honeybees are a great example of this—but animals who have this old-mammalian part in their brains possess the ability to live in incredibly complex social structures with systems of

---

[6] I fully acknowledge that this is a simple way of looking at the most complex organ that exists in any animal that we know of. Yet, since this book is not a scientific textbook, I believe Dowd's illustrations are highly applicable for the conversation at hand.

altruism and interchange that are noticeable in behavioral traits. Take for instance the highly complex meerkat social structure. These cuddly little mammals have the kind of social hierarchy that is found in human societies. Some meerkats act as leaders of the group while others function as sentinels whose sole job is to sit high up in a tree to watch out for eagles looking to prey on the foraging group. The development of the old-mammalian part of the brain was hugely important to this ability to be social and interactive.

As creatures evolved, the *new-mammalian* part, or the neocortex, developed and brought with it two major functions important for higher mammals such as ourselves and other primates: the ability to comprehend and to predict. This improvement was necessary for allowing the ability to rationalize and anticipate behavior in other creatures. What is more, the new-mammalian part of the brain also gave us the ability to self-deceive—a critical faculty for helping the old-mammalian and reptilian parts of the brain to get what they wanted: more food, sex, and a higher social status. Just as this new development was helpful for mammals, so also it had its downside: manipulation, coercion, and lying.

This might be a good place to mention that all of these parts of the brain have the capability to produce behavior that is morally neutral, such as the desire to have sex and eat; to morally good behavior, such as loyalty and fidelity; to morally reprehensible behavior, such as lust and gluttony. Loyalty can easily slip into cronyism and fidelity into an unhealthy obsession.

The last part of our brains to be developed—the *advanced* part—is the frontal lobe, or the prefrontal cortex. While it's the youngest part of the brain, it certainly shoulders the most weight for our personality—giving us much of our temperament and helping determine whether we make good or bad judgments. This part of the brain doesn't fully mature until humans are in their mid-twenties and, according to Dowd, this late development is something that car insurance companies are well aware of! Young men in particular—not least because of the extra testosterone processed in the reptilian part of their brains—often display risk-taking behavior devoid of the sorts of good judgments that come with a fully mature frontal lobe. I remember my parents being totally shocked at the price when my older brother had to get insurance as a teenager. It was astronomical, but for perfectly good reason. His frontal lobe at the time wasn't fully matured and likely couldn't handle the impulses generated within the reptilian part of his brain. The insurance company was not willing to take the risk that he was any different than most seventeen-year-olds.

But for fully developed male and female adults, the advanced part of the brain can, and often does, override the reptilian, old-mammalian, and new-mammalian drives. I know it's difficult to do, but have you ever smelled bacon, felt the urge to consume it, and stopped yourself? If you're in a romantic relationship, in the name of fidelity to your significant other, have you ever walked by an attractive person and decided not to entertain sexual thoughts? Of course you have. We win and lose behavioral battles every day based on, in part, how well we encourage our prefrontal cortex to override our cerebellum.

This knowledge of how our brain works can be incredibly helpful in regard not only to addressing our moral behavior but also our Christian life. Our youngest son, who's four years old and a ball of energy, is my little buddy. But the different parts of his brain pull him in various directions. He's torn between his advanced brain and his reptilian brain. I came home from work the other day and he ran to the door to greet me, and I saw that he was holding a Lego airplane that he had spent most of the afternoon building. It was pretty neat, and he created it all by himself and pored over every detail. Not thirty seconds later, as I'm still talking to him about it in the entryway, his older brother comes by and grabs it out of his hands—a mean thing to do, sure, but typical older brother stuff. Well, the little guy totally *lost it*! He freaked out with intense fervor and ran upstairs to start his reign of terror on his brother's room. He was right to be upset. He was protecting the thing that he worked hard on. The reptilian part of his brain was responding to a major threat. At first, I was rather upset at the situation. But when I stopped and thought through why he was upset—the wave of emotion from eons of needing to protect food or tools in order to survive—it made his actions much more understandable. More importantly, it provided me an opportunity to teach him, and myself, why we feel these emotions and that it's okay to feel them. It's not really okay to destroy stuff—we talked about that too—but it is okay to experience frustration and even rage. Step one is understanding where our emotions come from; step two is learning how to resist the temptation to let those emotions take control of us. To fail to do so can lead to sin. The kind of person we are developing into necessarily depends on what we do with those feelings.

TAMPING RODS AND TABLOIDS: OUR FUTURE CAN CHANGE

The famous nineteenth-century railroad worker Phineas Gage was a foreman working on a railroad in the mountains of Vermont when, in 1848, a disaster unfolded. While leading a group of men who were blasting away rock to make way for the rail, Mr. Gage was accidentally struck in the face with an iron tamping rod the size of a javelin. By all accounts, explosive powder was put in a small hole in order to clear earth for the rail. The hole should have been covered up with sand, which would have sent the blasts away from the workers, but someone forgot to do that. Gage's tamping rod accidentally struck an adjacent rock and the spark ignited the powder, sending the three-foot-seven-inch rod cleanly through Phineas's skull like a missile, landing eighty feet down the way. The remarkable part about this story is that Mr. Gage didn't die; in fact, he lived another twelve years! To the shock and wonder of the doctors and surgeons who cared for Gage, he survived without part of his prefrontal cortex and even went on to lead an active life.

But that's not the end of the story. After the traumatic brain injury, Mr. Gage's personality changed. The once mild-mannered, quiet man expressed different personality traits after the accident: a quick temper, vulgar language, and possible substance abuse. Thus, for some years after this incident, Gage struggled with the ability to make healthy decisions. Yet, despite such struggles, after almost a decade of routine and training, Mr. Gage was finally able to relearn how to control his behavior. To put it in evolutionary terms, Mr. Gage's advanced brain lost the ability to control his reptilian and old-mammalian brain. But here's the hopeful lesson that's incredibly important for our discussion: after years of working to rebuild those cerebral connections, Mr. Gage was able to regain some control over his more primal instincts.

As Christians who are open to learning from our evolutionary ancestors, the Phineas Gage story is a mix of warning and hopefulness. What happens to our brains actually matters. The chemicals that we put in our bodies that negatively impact our brain activity, permanently or temporarily, have a direct impact on our personalities and behavior. And, from our Christian point of view, altering our brains has direct correlation to our ability to develop as faithful and holy people. By teaching our children, friends, loved ones, and neighbors about the importance of both taking control of our advanced brain and protecting our advanced brain, they can be better prepared for life's complicated circumstances and temptations. Teenagers who know why their stomachs are in knots when they see a sensual advertisement on TV or stumble onto an explicit website are better able to acknowledge where the emotion is coming from and not let the wrong part of the brain take control. People who struggle with their weight can be more prepared against the supernormal stimuli of Burger King french fries when they remember that they don't live in the Stone Age and need to store fat on their waists for winter. It might sound a bit cliché, but knowledge really can be a powerful tool to shape our moral behavior.

Do you remember the study at the beginning of the book about the marshmallows and the children? The Phineas Gage incident is not unlike Walter Mischel's Stanford study because they both teach the same lesson: behavioral traits can be nurtured. It likewise brings hope to my young son, who was mesmerized by the tabloid with the picture of the woman on it. We have a chance at reordering which part of the brain will have ultimate dominance—whether we're people who have purposely altered our brains via addiction or whether we're children who have impatiently grabbed for a marshmallow or who refuse to look away from a tabloid. There's incredible hope in this notion. Yes, Mischel is right: children that show they can delay gratification at a young age will fare better in life. And, yes, the surgeons were right: Phineas Gage's personality and decision-making capabilities were negatively altered after his brain was injured. But both show how rehabilitation and reformation can take place in an individual. Children can be trained to delay gratification and have deeper self-control. Mischel and his group have worked with *Sesame Street* and other outreach programs to demonstrate this truth. Through the correct regimented social circumstances, Gage was able to incrementally mitigate his temper and mostly regain the person he was before the accident. If we can learn anything from these incidents, it is that we can know that change and development are in fact possible. As we will see, Christians are in the perfect position to positively nurture what the process of evolution has given us through the ages.

7

HAVING EVOLUTIONARY ROOTS ISN'T JUST BAGGAGE

Pursuing a Holistic Understanding of Redemption

F OR THE SAKE OF THE CHURCH it is imperative that we move away from dualist thinking that has crept into the church over the generations. We are *whole* people—physical, mental, intellectual, emotional, and spiritual beings. We're not merely products of our experiences and mental states, nor are we simply spirits stuck in our bodies. Any notion to the contrary is a reductionism that not only goes directly against Christian tradition and Scripture but also our own reason and our experience of how life operates.

I'm sure that I'm the only one to experience this in my teenage years, but I distinctly remember fighting with my parents when it was so obvious that it was *all their fault*! Just imagine a scrawny thirteen-year-old waiving a defiant finger in the air, saying, "It's either nature or nurture, but either way it's your fault, Mom and Dad!" As brave as those outrageous statements seem to a teenager—and as ludicrous as they are in reality—my parents were less impressed and weren't caught off-guard. I can't quite remember their response, but I'm sure it was something along the lines of "I brought you into this world, and I can take you out of it!" They were right to object to my teenage abdication of responsibility. Blaming my poor behavior on my nature and their nurture (i.e., on genes and environment) is tantamount to saying, "The devil made me do it." From a Christian perspective we are whole people with natural and environmental influences, conviction from the Holy Spirit, and the determination of our own wills that shape not only our actions but also our personalities and identities.[1]

This might come as a surprise to you, but many of our Sunday school classes are more influenced by Plato than they are by the Bible. As numerous scholars have pointed out, Platonic thought, which advocates for a clear separation between the body and soul, weaseled its way into the folk theology of many Christians.[2] For a general introduction to this, notice what theologian Randy Maddox observes: "Some Greek portrayals of the body/soul relationship assigned the body a primitive, if not actively antagonistic, impact on spiritual life. By contrast, the Bible presents the body as part of God's original good creation, and sin as a distortion of every dimension of human life."[3] In other words, our actions are not just the spiritual struggles of a spiritual being trapped in a fleshly body. Instead, both our sins and our holy actions should be seen in the light of the whole person—a nonreductive person who is a complex agent with a physical, emotional, spiritual, intellectual, and mental side.[4]

Finding common ground between our contemporary understanding of evolutionary science and both historical and contemporary Christian theology allows us to fully embrace the unity of the

---

[1] The Holy Spirit initiates the process of becoming holy as well as guides us on that journey.
[2] See N. T. Wright, *Surprised by Hope: Rethinking Heaven, the Resurrection, and the Mission of the Church* (New York: HarperOne, 2008), 80.
[3] Randy L. Maddox, *Responsible Grace: John Wesley's Practical Theology* (Nashville: Kingswood Books, 1994), 72.
[4] Philosophers and theologians call the body-soul synthesis "nonreductive physicalism." See Joel B. Green, *What About the Soul? Neuroscience and Christian Anthropology* (Nashville: Abingdon Press, 2004). See also Joel B. Green, *Body, Soul, and Human Life: The Nature of Humanity in the Bible*, Studies in Theological Interpretation (Grand Rapids: Baker Academic, 2008).

whole person instead of unhealthy dualisms that grow out of veiled Platonic thought. In a very real sense, unifying science and Christian theology helps us focus on the important physical reality of the Christian faith: the incarnation, the virgin birth, the physical resurrection of Jesus, and the physical resurrection of humankind. By drawing attention to these important physical doctrines, we rightly call to mind the physical reality of our Christian living.

In fact, most of Paul's writings—which make up the majority of the New Testament—are an attempt to combat dualistic Platonic (or gnostic) thinking. So we're keeping good company when we articulate the importance of the whole person rather than believing in the notion of an anemic soul-only self. After we come to grips with this important theological distinction, we can see how important understanding our evolutionary roots is to our Christian faith. We can start to notice where our urges come from and what our temptations are rooted in—and then after we notice them, we can be intentional to change our behavior. As we discussed earlier, knowing *why* we want to eat five Big Macs and stare at an attractive person can be advantageous in tempering such desires.

A "GLASS HALF FULL" PERSPECTIVE

When it comes to our biological makeup, it's not all doom and gloom. In fact, the notion that our evolutionary roots have made humans to be entirely selfish has been largely argued against and disproved in the last decade. For some time before this new understanding, altruism—that is, acts of kindness and selflessness—was considered a taboo idea within the field of evolutionary biology. How could an organism that acted altruistically, going against its own needs in order to help others, end up being more fit for reproduction? Therefore, so the thought went, the survival of the fittest meant that organisms have to look out for number one. However, there was one major problem with that line of thinking: organisms don't live in isolation; they typically live in groups. And it goes without saying that groups that cooperate are healthier (i.e., out reproduce) than groups that are uncooperative.[5]

To use the language of evolutionary biologists, organisms came about in order to act as hosts for replicator genes. In a rapid-changing environment, these organisms—forced to compete with harsh consequences where only those best adapted to the particular environment survive—had to find means of adapting faster than other entities less able to adapt. Cooperation in groups, then, arose because it gave an adaptive advantage.[6] It's the same concept as to why people save for retirement. A little bit of suffering now will be better for my family—and future generations—in the future. So, as eons passed, genes in cooperative groups were selected to sacrifice in the short term in order to benefit their individual gene fitness (via the group) in the long term. The payoff in the end was greater than unadulterated selfishness, and a form of altruism had begun.

Philosopher of science Michael Ruse approaches the altruism phenomenon by arguing that kin altruism (acts of kindness toward relatives) and reciprocal altruism (acts of kindness for some kind of payback) explain the bulk of altruistic acts. For Ruse, there is a connection between a sense of sympathy and likelihood of reciprocity.

---

[5]For those who are in field and might object to such "group selection" or multilevel selection, I write about this in my previous IVP book, *Evolution and Holiness* (Matthew Nelson Hill, *Evolution and Holiness: Sociobiology, Altruism, and the Quest for Wesleyan Perfection*, Strategic Initiatives in Evangelical Theology [Downers Grove, IL: IVP Academic, 2016]). Some of this section is used by permission from that book. See also Edward O. Wilson, *The Social Conquest of Earth* (New York: Liveright, 2012), which is a great overview of the new research in evolutionary altruism.

[6]Cooperation is always found in a benefits-to-cost ratio, whether knowingly or not. See Martin A. Nowak, "Five Rules for the Evolution of Cooperation," in *Evolution, Games, and God: The Principle of Cooperation*, ed. Martin A. Nowak and Sarah Coakley (Cambridge, MA: Harvard University Press, 2013), 109. Altruism is sometimes thought to be a subset of cooperation. "The form of psychological cooperation called altruism . . . is a subset of evolutionary cooperation. Cooperative behaviors count as altruistic only when carried out by an agent when (1) she possesses cognitive, affective, and dispositional states of sufficient complexity that they can count as motivating her actions, and (2) at least some of her actions are motivated by 'goodwill' or 'love for another'" (Philip Clayton, "Evolution, Altruism, and God: Why the Levels of Emergent Complexity Matter," in Nowak and Coakley, *Evolution, Games, and God*, 346).

Organisms have evolved to where they may be dependent on one another through social bonds—such as flock, herd, or other community—or are at least organisms that are capable of reciprocating good deeds.[7] While many acknowledge that kin and reciprocal altruism are the dominant forms of altruism, they do not *solely* explain the phenomenon.[8] Turning to a theologian in our own faith, Thomas Aquinas also saw kin and reciprocal altruism as the basis for much of human behavior in the natural instincts that humans shared with other animals: reproductive behavior, care for kin, and the like. But it is the *moral* sanctioning and forbidding of certain activities that make humans fundamentally different and unique from the rest of the animal kingdom.[9] The other day my five-year-old son was asking me if it was wrong for a lion to kill another lion. He said, "Dad, are there lion police officers and lion judges?"

"No buddy, they don't have the same kind of society as we do," I responded.

"But the lion is the king of the jungle, right? They must be able to rule on things like that!"

Fair point, son. Fair point. But if we might be able to move beyond this level of thinking, we'd quickly find that Aquinas is correct and that

---

[7] For further reading on this concept see Stephen G. Post et al., eds., *Altruism and Altruistic Love: Science, Philosophy, and Religion in Dialogue* (New York: Oxford University Press, 2002), 168-81. Here, Pope goes into detail on this idea in his section titled "Relating Self, Others, and Sacrifice in the Ordering of Love."

[8] Darwinism was originally thought to show that human nature was completely individualistic and selfish. Altruism, then, was regarded as the product of culture/environment alone, with "pure altruism" being unattainable. A better way to deal with altruism, however, is to provide awareness that the error of sociobiological fatalism does not rest in recognizing biological causality but through minimizing the force of a multitude of other causal factors such as personal, cultural, economic, and so on. Here, human motivation in particular has multiple interacting causes that impact decision-making, altruism, or other subjective or moral actions. For more on this, see Stephen J. Pope, *Human Evolution and Christian Ethics*, New Studies in Christian Ethics (Cambridge: Cambridge University Press, 2007), 214. See also Stephen J. Pope, *The Evolution of Altruism and the Ordering of Love*, Moral Traditions and Moral Arguments (Washington, DC: Georgetown University Press, 1994), 105.

[9] Craig A. Boyd, "Thomistic Natural Law and the Limits of Evolutionary Psychology," in *Evolution and Ethics: Human Morality in Biological and Religious Perspective*, ed. Philip Clayton and Jeffrey Schloss (Grand Rapids: Eerdmans, 2004), 228.

humans are unique in our moral sanctioning. And if *sanctioning* and *forbidding* are special to the human race and are part of what makes us human, then we must conclude that some kind of pure altruism is at the very least possible.

For the sake of the church it is also important to recognize that humans have altruistic tendencies that developed *during* evolution. Very recently in the history of studying evolution, altruism has been studied by numerous scientists who appeal to genetic and evolutionary explanations of human behavior. Within the field of biology, fresh studies by primatologist Frans de Waal, among others, suggest that humans are *not* mostly selfish (as has previously been argued in this field); instead, as de Waal demonstrates, the biological makeup of humans have *biological* capacities for altruism.

A little history on the roots of these discoveries might help you understand why this discovery is revolutionary for the field of sociobiology or the field that studies evolutionary roots to animal (including human) behavior. What excites me the most is that these discoveries line up with what Christians have been arguing for the past two millennia: namely, that in addition to having a penchant toward maleficent behavior, humankind naturally has a great capacity for goodness.

Frans de Waal, in particular, has numerous studies that display how primates (some of whom are our closest nonhuman ancestors) have a biological bent toward kindness, empathy, and altruism.[10] Take for example a study examined in de Waal's *The Age of Empathy* that connects human behavior to primate behavior, displaying the connection between innate capacities for altruism.[11] In this study, a capuchin monkey

---

[10]Frans de Waal has numerous works relating to this. A prominent one in particular is Frans de Waal, *The Age of Empathy: Nature's Lessons for a Kinder Society* (New York: Harmony Books, 2009). Another good biological account that serves as an alternative to Dawkins is Kenneth M. Weiss and Anne V. Buchanan, *The Mermaid's Tale: Four Billion Years of Cooperation in the Making of Living Things* (Cambridge, MA: Harvard University Press, 2009).

[11]This and the next two paragraphs are lightly adapted from Matthew Nelson Hill, *Evolution and Holiness: Sociobiology, Altruism, and the Quest for Wesleyan Perfection*, Strategic Initiatives in Evangelical Theology (Downers Grove, IL: IVP Academic, 2016).

reaches through an armhole to choose between two differently marked tokens, while another monkey, physically separated from the first, looks on. The tokens can be exchanged for food, but in different ways. One token feeds both monkeys, and the other token feeds only the chooser. Capuchins typically prefer the more prosocial token.[12]

So, is this an altruistic action, since it involves no direct reciprocation and offers no benefit to the altruistic agent? If so, is that altruistic action in contradiction to the idea of "being born selfish"? Here, sociobiology is functioning merely through evolutionary biological explanations applied to the level of not only physical traits but also behavioral traits, showing that there *is* such a thing as Darwinian altruism (dying for the hive, etc.). Through this kind of altruism that benefits the agent, sociobiologists might try to explain the concepts of egoism and altruism on the biological level but are unable to do so entirely or completely due to the existence of either non-kin or non-reciprocating (i.e., pure) altruism.

Another study done at Taï National Park, Côte d'Ivoire, showed that chimpanzees took care of those in the group they purposefully lived with. When a leopard injured some group members, others licked injured chimpanzees' wounds to remove dirt and waved flies away from the infected areas. They were also mindful of the injured members by slowing down the travel speed in order to keep them with the group.[13] This purposeful group behavior, even when it put the lives of the healthy in danger, made sense when looking at the group benefit. However, because group members functioned more efficiently and safely as a whole does not mean that there was not an element of selflessness and sacrifice to stay in the group. There was an opportunity for the chimpanzees to cut their losses, especially when given the danger of caring for the wounded. Instead, they chose to act altruistically.

---

[12] de Waal, *Age of Empathy*, 194.
[13] de Waal, *Age of Empathy*, 7.

Consequently, sociobiologists can provide grounds for a view of human nature that is not solely egoistic but takes into account genuinely altruistic, as well as selfish, motivations.[14] The conversation does not have to be limited to biological influences on behavior. Environment, then, can play a critical role in the evolutionary process. This is the old nature-versus-nurture question. The fact that the environment by itself, on certain occasions, can provide the ingredients required by the process of natural selection gives environmental influences the status that critics of biological determinism have championed.[15]

Elliot Sober and David Sloan Wilson describe how a Sudanese tribe of people called the Nuer, dominant over a century ago, had a complicated and costly social ordering system that bonded the group together.[16] The group closeness and structure in turn made them more dominant in the region, especially in regard to military campaigns. This work retrieves aspects of evolutionary thought that have fallen from grace with the preeminence of individual selection and gene-level selection in neo-Darwinism.[17] Sober and Wilson claim that it is having to somehow find a middle way where free will acts as the deciding factor in how organisms react to their genes and environment—choosing either to act altruistically or selfishly.

The authors also claim that "natural selection based on cultural variation has produced adaptations that have nothing to do with genes."[18] In the same vein Holmes Rolston suggests about group selection that "tribes of Good Samaritans will out reproduce tribes of thieves."[19] Rolston later argues that theories of altruism based on the

---

[14]Pope, *Evolution of Altruism and the Ordering of Love*, 110.
[15]Elliott Sober and David Sloan Wilson, *Unto Others: The Evolution and Psychology of Unselfish Behavior* (Cambridge, MA: Harvard University Press, 1998), 337.
[16]Sober and Wilson, *Unto Others*, 186-91.
[17]Pope, *Human Evolution and Christian Ethics*, 220.
[18]Sober and Wilson, *Unto Others*, 337.
[19]Homes Rolston, "The Good Samaritan and His Genes," in Clayton and Schloss, *Evolution and Ethics*, 244.

genetic transmission of behaviors, and particularly the constraints on altruism that come about in these contexts, no longer hold when altruistic values are culturally or religiously passed on.[20] In other words, he's expressing that, in addition to passing on genes of kindness, altruism and acts of kindness can be taught and environmentally passed on to the next generation.

These examples of nature and nurture influences on an organism, coupled with the ability to freely act selfishly or altruistically, lead us to be more balanced on the selfish-selfless spectrum. In other words, humans have the ability to travel along this spectrum, moving between selfish and selfless behaviors. And if such movement is possible, intentional community that is structured by environmental constraints on behavior might draw humans toward altruism. Essentially: when we learn better, we do better and teach that behavior to our children. Yes, our behavior, whether sinful or holy, is in many ways influenced by biological urgings, but humans are uniquely positioned to be able to encourage positive behavior. Sin, then, is not inevitable.[21] Of course, we have debauched urges and behavior that stem from our biology; we need only to look at violence, torture, and manipulation (to name a few). But we also have positive and prosocial traits that are rooted in our biology. Just look at the way your face lights up when you see a puppy, or how you are overtaken by compassion when a loved one, or even a stranger, is hurt in front of you.

We are a conflicted people who navigate the full spectrum of selfless and selfish desires and behaviors. The other day I wanted to do something kind for my wife, so I said I'd take care of the kids in the morning so she could stay in bed to read. My kids were particularly crazy that

---

[20]Rolston, "Good Samaritan and His Genes." See also Philip Clayton, "Biology and Purpose: Altruism, Morality, and Human Nature in Evolutionary Perspective," in Clayton and Schloss, *Evolution and Ethics*, 319."

[21]The idea that sin is not inevitable has been in the Christian tradition for a long time. One of the main advocates for it, Gregory of Nyssa, who helped form the trinitarian doctrine, argued that the sinless life that progresses toward holiness is certainly possible. See Gregory of Nyssa, *The Life of Moses*.

morning, and so shortly after they awoke I went back into the bedroom and asked my wife, "Have you seen the mother of my children?"

She wryly responded, "Not yet I haven't!"

Touché! It hit me immediately that my altruistic tendencies had quickly turned selfish. Often, my "I'll do something nice for *someone else*" quickly turns into "Why doesn't someone do something nice for *me*?"

Remember, we are whole people: spiritual and physical. And in our Christian lives, our actions—be they for good or ill—are often the fruit of our living. Speaking of the negative aspect of our behavior, the early twentieth-century theologian Mildred Bangs Wynkoop puts it this way: "The body is not sin-bearing; it is basically good. Sin is an attitude and spirit of rebellion against God, not a substance."[22]

I also want to make clear something so we don't get tripped up theologically. To be sure, the relationship between morality and holiness can sometimes seem blurry. While I recognize the differences between acts of altruism, morality, and the spiritual process of becoming holy, these are not unrelated activities. In other words, being a moral person is not identical to being on the process of becoming holy, although they are related. One can be a moral person and not a holy person, but one can never be a holy person without being a moral person. We can take a cue from the book of James, where he connects works and faith: altruism and morality can be linked with holiness. One cannot become holy without producing the fruit of altruism and moral development. Similarly, those practicing altruism and positive moral behavior are in some sense heading in the direction of holiness. This process does not happen without the intervention of God. We can see an analogy of this process in the life of John Wesley, who, when struggling with his faith, was exhorted by his friend Peter Boehler, "Preach faith till you have it and then because you have it, you will

---

[22]Mildred Bangs Wynkoop, *A Theology of Love: The Dynamic of Wesleyanism* (Kansas City, MO: Beacon Hill Press, 1972), 49.

preach faith."[23] In a similar way, while practicing altruism is not the sole path to holiness, they are very much linked to one another.

CULTIVATING HOLINESS

We are used to hearing about the miracle of a baby, but let's talk about the miracle of the *mother* for a moment! Now, I'm not trying to score points with my wife—I'd never do a thing like that—but I remember staring at her shortly after the birth of our first child and being struck by how amazing and miraculous the human body is. She, with very little help from me, grew an entire human inside her body. She carried this child around in her womb for months, providing all the nutrients and safety needed to form a healthy baby. Then, in courageous fashion, she birthed this little human out of her body. And if this all weren't astonishing enough, she immediately started feeding this new human *with her own* body. Simply miraculous! During the same period of time, on the other hand, I was able to watch a baseball game throughout our hospital stay. Not so miraculous.

I think most Christians would agree that in a very real way, creating, birthing, and feeding a baby is simultaneously the most miraculous *and* the most natural thing in the world. Every day, millions upon millions of women all over the world go through this same process. Right now there are countless mothers who are either pregnant, birthing a

---

[23]See Hill, *Evolution and Holiness*. See also J. Wesley and A. C. Outler, *John Wesley* (Oxford University Press, USA, 1980), 17. In *After Virtue*, Alasdair MacIntyre expresses a similar notion by using virtue language: "The immediate outcome of the exercise of a virtue is a choice which issues in right action: 'It is the correctness of the end of the purposive choice of which virtue is the cause' (1228a1, Kenny's translation, Kenny 1978) wrote Aristotle in the *Eudemian Ethics*. It does not of course follow that in the absence of the relevant virtue a right action may not be done. To understand why, consider Aristotle's answer to the question: what would someone be like who lacked to some large degree an adequate training in the virtues of character? In part this would depend on his natural traits and talents; some individuals have an inherited natural disposition to do on occasion what a particular virtue requires." See Alasdair C. MacIntyre, *After Virtue: A Study in Moral Theory*, 3rd ed. (Notre Dame, IN: University of Notre Dame Press, 2007), 149.

baby, or nursing a child. It's very natural and very normal. Yet, the created life—even the process itself—is also strikingly miraculous.

If we can understand this concept, it's not a far stretch to notice the both-and progression of how we move toward holiness, or what theologians call the process of sanctification. It's a both-and undertaking that includes both God and his follower: God's movement toward a person followed by a freewill movement back to God. It takes both God's action (which is miraculous) and our action (which is natural) to move toward holiness. As mentioned earlier, this natural and supernatural endeavor rests on nondualistic thinking about our bodies and our spirituality. We're not simply ghosts functioning in some machine but rather whole persons moving with the Holy Spirit's prompting and leading.

Through a combination of human choice and the grace of God, humans possess the ability to continually *overcome* their genes and live holy lives.[24] Such grace is much the same as what theologians call "prevenient" or "common" grace as it works alongside the theological concepts of original sin, justifying, and regenerative grace. It also aids believers during and after the free conversion and works alongside sanctifying grace as they move toward living holy lives. With the use of *overcome*, I mean to convey the connotation that grace helps us go

---

[24]When using the word *overcome* I always mean to confer the sense of the "continual process of overcoming." By doing so, I hope to draw attention to both the present progressive *and* future progressive nuances of the verb *overcoming*. At the same time, I do not want to mitigate the present perfect nature of the word *overcome*, which helps us understand the grace that helps the individual overcome biological proclivities toward violence, for instance, both now and progressively in the future. Thus, owing to its perspectival nuances, I have chosen to employ the word *overcome* to describe how a Christian can become holy *in this life* without being determined by their biology; for I believe this process embodies the heart of Christian holiness. A holy Christian is not uninfluenced by biology but, through the working of grace, is able to overcome the limits of biological influences. I have chosen the word *overcome* instead of *transcend* because I do not wish to convey the concept of otherworldliness or to fall into gnosticism. I also do not wish to convey the idea that the human has become something other than fully human. It is my argument, then, that the word *overcome* expresses the truth that, by the human will and the grace of God, we are not bound by genetic proclivities. I remind the reader that much of the ideas in this section and chapter are taken from my book *Evolution and Holiness*. While much more academic in nature, the reader who wants to go deeper might be interested in exploring here.

beyond the limitations and propensities of our environmental and evolutionary influences, and grace allows us to continually overcome the constant proclivity toward self-indulgence and, ultimately, sin. Through this process, human actions and choices are not determined by innate tendencies, passions, or biological predispositions.[25] Yet the genetic urges pushing us toward the fifth cheeseburger or pornography do not dissipate or go away; we merely have the grace to overcome the genetic urges.

An individual's works are always a *response* to grace. We are never able to work hard enough to attain either what theologians call justification—that is, a state of rightness with God—or holiness. Instead, just as prevenient and common grace beckon individuals toward a justifying Savior, so does this sanctifying grace call the Christian to respond to the Holy Spirit's urgings. At some point the individual becomes more influenced by the Holy Spirit's persuasions than by evolutionary proclivities, moving them further along the path of sanctification and ultimately enabling the individual to overcome those genetic and environmental constraints. To be clear, this is not some kind of dualism of the spiritual-material but rather a portrait of God working *within* creation through grace, allowing humans freely to respond.

We might liken this phenomenon to having God on a rope.[26] Imagine the moment of justification as the moment when the Christian becomes connected to God via a rope, which signifies grace. Through the Holy Spirit's urgings, the Christian can draw the rope inward, shortening the distance between the Christian and God, closing the distance between where they are and holy living. Since the rope is always connected between the individual and God, the Christian can let out the rope—freely moving away from holy habits and rejecting

---

[25]To be clear, if it were the case that genes totally determined actions without free will, rather than merely predisposing actions, it would skew what we think about sin. Yet there is currently no conclusive scientific reason to believe that genes totally determine actions.
[26]Again, this is an analogy used in *Evolution and Holiness*.

God's grace—or take in the rope.[27] At some point, however, the Christian is so closely connected to God that the rope does not have to connect the two. A Christian's works, then, are always a *response* to grace; a habit of holy living is unable to be earned when divorced from God's activity.

Another way to think of how God's grace works with human freedom to allow someone to overcome genetic urges and constraints is the process we undertake when attempting to fly a kite. Flying a kite requires an incredible amount of activity: putting together the kite, clearing space for running and flying, checking for imperfections in the tail or the kite itself, and so on. Yet, regardless of how much work the individual does to the kite to prepare it to fly, it is all useless—in fact impossible—to do without wind. The wind is the active agent needed for the kite to fly. In the same way that preparing a kite for flight only makes sense in relationship to the wind, and so are all the good works of a Christian only intelligible if they are preparing to be directed by God. God is the active agent through his free grace.

Factoring in the Holy Spirit's action and our free will, we can see that the concept of holy living can be compatible with current knowledge of evolutionary biology if we are to assume that God works with the Christian not to replace or work in parallel with genetic material but rather to work within the total human context by providing the grace by which a Christian can overcome negative evolutionary and environmental constraints.

---

[27] I have found this analogy, while unhelpful to some, to also be very helpful especially when trying to explain how a Christian might be able to "throw away" salvation but not "lose it." For instance, after justifying grace, the individual is connected with God. One may let out the rope to an extended distance but is still attached, and thus not cannot lose salvation (as we might lose a wallet or car keys). At some point, however, the individual might choose to untie oneself from God and throw the rope aside.

NURTURING NATURAL VIRTUES TOWARD HEALTHY COMMUNITY

THE DENTIST TOLD MY OLDEST SON that he needed braces. My son bravely sat through the talk as the dentist described how she would pull a couple of teeth to make room for the movement. When we got home, my son said that he was anxious about the procedure, and I assured him that they would use some laughing gas, which would make him feel happy and hysterical. Somehow my shameless reassurance of drug-induced bliss didn't much comfort him, and he still seemed petrified. I then explained to him that he would need to go to the dentist every week for ten weeks in the beginning. This made him all the more nervous, and I couldn't quite understand why.

That night when I was tucking him in, he said something that I'll never forget: "Dad, if you think this is best for me, then I'll do it."

"Wow, Buddy," I said, truly touched by his trust in me, "that's a really mature thing to say."

"Dad, I was just so nervous because it seems like a lot of weeks of pulling teeth, but if you say so, I'll do it."

"Buddy, you just get teeth pulled one time. What were you thinking would happen?"

"Oh! I thought I would get several teeth pulled every week for ten weeks, and I kept wondering what teeth would actually be left to put braces on!"

Though at that moment I felt terrible for not realizing sooner why he was so afraid, upon reflection I am moved by his trust and obedience—traits he must get from his mother. He was truly terrified, yet he trusted. He felt every urge in his body resisting what was being asked of him, yet he trusted his father—wrongly in this instance; he thought that he would get every tooth in his mouth pulled over a series of ten torturous weeks. Despite those fears and instincts, he overcame them and put his trust in something that was beyond him.

Our Christian experience is like this childlike trust—we're often asked to trust the guidance of the Holy Spirit and those in our community of faith, even when it is difficult, scary, or confusing. If, as I have argued in this book, one of our spiritual tasks is overcoming our genetic and environmental drives, then this work cannot be done in isolation. If we embrace the truths of evolution, we are making but one small (though important) step toward recognizing which areas of our lives need nurturing and attention. Awareness of the biological forces at work in us equips us to address the origins of our struggles. If I succumb to the desires of a lust-filled mind, then understanding where those desires come from can help me begin to find accountability. If giving in to overeating feels like a cycle I'm simply unable to break, then going to the root of the problem should be my starting place. And if I have a proclivity toward addiction, quick-tempered behavior, or impatience, then I can go to the beginning and consider the evolutionary seeds of such behaviors.

This is not to say that embracing evolution will solely fix these problems, but it's a start. To this end, I am dedicating space in this final chapter to offer practical ways through which we can begin the process of overcoming negative genetic and environmental urges that push us away from holy Christian living. This process of overcoming cannot be done in isolation—perhaps a contrary thought in a Christian culture that often sees spiritual formation as a process that involves only the person and Christ. Here, John Wesley, the eighteenth-century father of Methodism, will prove to be prophetic in his own writings and teachings on moral and spiritual formation. For, according to Wesley, "There is no holiness but social holiness," that is, doing Christian life in community.[1] This claim is profound. It may also discomfort us a bit. Yet it is compatible with evolution and the Christian faith: we can only develop spiritually together, in community, with the Holy Spirit and with other believers.

GETTING THE RIGHT KIND OF COMMUNITY

In previous chapters we noted that environmental constraints have the potential to temper the biological constraints placed on humans from birth. For instance, each human being is born with their own set of genetic influences that modify and guide behavior, such as their propensity toward violence or lust. Yet, as Christians have modeled throughout the millennia, living in a structured community—full of shared experiences and accountability—can help us overcome the negative biological urges and, instead, nurture positive traits inherited from evolution. This shared experience and influence is part of the reason why we have churches. In addition to receiving the sacraments and worship, we have a unique opportunity to influence others toward holiness. Going to church isn't always about us "getting fed," but rather about how we can influence others toward holy living.

Humans, of course, do not live in isolation and are influenced by their environments, which include other people. Community has always been part of the context humans live in. So it is no coincidence

---

[1] See the whole quote from Wesley: "The gospel of Christ knows of no religion, but social; no holiness but social holiness. Faith working by love, is the length and breadth and depth and height of Christian perfection" (John Wesley and Charles Wesley, *Hymns and Sacred Poems* [Bristol, UK: F. Farley, 1739]).

that, although evolutionary theory claims that it is difficult to learn how to become better (more holy) individuals, evolved traits are always trained within particular communities.[2] Consequently, each community has the potential to push the boundaries of biological constraints wider and wider, and this can be both positive (and negative) for Christian individuals. We humans are amazing at adapting to our surroundings. In the same way, when we dwell among people that are Christlike or otherwise, we tend to adapt to the standards of that community. It might sound like common sense, but we become like those we surround ourselves with.

When discussing the kind of community where individuals might be encouraged to live holier lives, it's important to remember that many Christians, while claiming to be faithful, live lives that bear little resemblance to the high ethic of the New Testament.[3] Thus, it's imperative that we surround ourselves with not just any Christians but the *right kind* of Christians. Give careful thought to these words from theologian and philosopher Stephen Pope:

> Because of the complexity of history, Christians have been both slave owners and abolitionists, soldiers and pacifists, pro-choice and pro-life. Christians are well prepared to acknowledge the historical limitations imposed on their own judgments. This is precisely the major weakness of the evolutionists—the failure adequately to acknowledge the historical and cultural nature of the human being. Whatever universal species-specific biological traits we have will always bear their moral significance within particular cultural contexts.[4]

---

[2]Take for example, human sexuality. Though this is a natural trait, it is conditioned by a host of norms from specific communities. See Stephen J. Pope, *Human Evolution and Christian Ethics*, New Studies in Christian Ethics (Cambridge: Cambridge University Press, 2007), 226-27.
[3]Pope, *Human Evolution and Christian Ethics*, 223.
[4]Pope, *Human Evolution and Christian Ethics*, 234.

Even given evolution, we can still be influenced by the environment we reside in. This reality is why dwelling in a particular community might be necessary for Christians to develop a lifestyle of holiness. I realize that this idea might seem tedious, but any community that is the *right kind* of community—one that can nurture a lifestyle of holiness—is going to be purposeful and, more likely than not, ordered.

A few years ago I wanted to learn how to keep bees and met a commercial beekeeper who was willing to show me the ropes. He would let me follow along with him a few times a week over the summer, and I would "lift stuff," which was about all I was good for. Every few days I would spend time with him and other beekeepers. It was hard and hot work, and I got stung more times than I can recall (even a few times running away, screaming!). But the group of beekeepers would encourage me to get back to work because that's what was needed for me to become a better beekeeper. After a summer of countless hours working on this commercial farm, I began to see the ordered work paying off. Not only did I know new skills such as how to calm an ornery hive or how to extract honey, but I also trained my chicken self not to be so intimidated and nervous. The routine of working with bees led to the formation of habit in me—I was changed because of this ordered work. But the change was not simply in my knowledge of beekeeping but also in my character.

Any kind of ordered and accountable community—whether concerned with beekeeping or character formation—doesn't happen in isolation. Social holiness, that is, the virtue of working with others toward overcoming predispositions inherited through evolution, needs to be at the heart of healthy Christian communities. Together, we can learn techniques that are distinctively worked out through a kind of purposeful culture nurtured in small Christian communities.

Such a cooperative and ordered community comes from both the grace of God and the grace of the group. As Thomas Oden articulates, "When we cooperate with the unmerited grace of God's saving act on the cross, we [should] not forget that it is precisely grace that *enables* our cooperation."⁵ Even if we are not part of the Wesleyan tradition, we can agree that John Wesley gives us an example of just the sort of ordered community wherein grace *enables* our cooperation.⁶ Wesley shaped what I call a *world of constraints*—an environment in which individuals were much more likely to move toward altruism and kindness than selfishness and cruelty, overcoming biological tendencies that pushed people away from holy living. It should be clearly stated that in most Christian theologies (Wesleyanism included), people don't do good work apart from grace.⁷ So the good work that one does within a community group is grace working with the human condition, not the human condition working autonomously and divorced from the grace of God.

In his teachings and practices John Wesley shaped a social holiness structure of constraints that pushed people toward altruism and habitual holy living. These constraints, which were the outcropping of his highly structured accountability groups, worked to suppress other biological tendencies that would have encouraged the individual to be particularly selfish. In this way, Wesley formed an organized community structure that moved underlying natural tendencies from self-interest toward altruism. Wesley's passion for change in people's everyday lives permeated all that he did. He felt a strong need to

---

⁵Thomas C. Oden, *John Wesley's Teachings: Christ and Salvation* (Grand Rapids: Zondervan, 2012), 2:152; emphasis mine. Oden goes on to say that "Though not intrinsic to freedom, grace is constantly present to freedom as an enabling, wooing gift. That does not reduce grace to an expression of nature. Grace remains grace. It is not something we possess by nature. It is given us. Yet grace is given abundantly to everyone, from the Paleolithic mound makers of Georgia to the forest Hottentots of Africa. Everywhere human beings exercise freedom, there grace is working to elicit, out of the distortions of fallen human nature, responses of faith, hope, and love. Preparatory grace remains a teaching that can be twisted so as to imagine that Wesley was covertly affirming the very Pelagianism he so frequently denied" (152). See also John Wesley, *Sermons 71-114*, ed. Albert C. Outler, Bicentennial Edition of the Works of John Wesley (Nashville: Abingdon, 1986), 3:201.
⁶For Wesley, exegesis and living are one and the same. See Carl Bangs, *Our Roots of Belief: Bible Faith and Faithful Theology* (Kansas City, MO: Beacon Hill Press, 1981), 41.
⁷Oden, *John Wesley's Teachings*, 2:141.

encourage modifications in the behavior of those who were ungenerous. We can easily track his encouragement in his sermons, especially those moments in which he employs fiery language when focusing on the seriousness of shifting one's disposition toward altruism and Christian love. To those who seemed predisposed toward altruism, Wesley likewise encouraged them through regular meetings, since he found accountability necessary to maintain changed behavior.

One shape this organized community structure took was the small group, which consisted of a handful of people to about twenty and had a rigorous set of constraints placed on the members by Wesley. Wesley called these small group *band-societies*, and he offered specific directions for how to care for each other.[8] For example, according to the "Rules of the Band-Societies," concerning the poor Wesley instructed the small groups "Zealously to maintain good works; in particular, 1) To give alms of such things as you possess, and that to the uttermost of your power. 2) To reprove all that sin in your sight, and that in love and meekness of wisdom. 3) To be patterns of diligence and frugality, of self-denial, and taking up the cross daily."[9] These instructions all pertained to outward holiness and helped the members be mindful of their actions. Yet these constraints were predicated on the discipline of inward holiness. Nonrhetorical questions concerning inward holiness were asked of those within a small community group: "1) What known sins have you committed since our last meeting? 2) What temptations have you met with? 3) How were you delivered? 4) What have you thought, said, or done, of which you doubt whether it be sin or not?"[10] Given that these were questions to be answered out loud within the band, Wesley's community groups were personal, requiring trust and honesty of all participants if those involved were to progress in both

---

[8] There is a full set of rules and directions for the "bands" in John Wesley and Rupert E. Davies, *The Methodist Societies: History, Nature, and Design*, Bicentennial Edition of the Works of John Wesley (Nashville: Abingdon, 1989), 77-79.
[9] John Wesley, *The Works of John Wesley*, ed. Thomas Jackson and Albert C. Outler (Grand Rapids: Zondervan, 1958), 8:274.
[10] Wesley, *Works of John Wesley*, 8:273.

their faith and actions. There was no room for static inward or outward holiness within this highly accountable social structure. One's faith was active and communal.

The hard work of these groups, which honed inward and outward holiness, was evident and plenteous. These eighteenth-century communities were offering their time and energy to a whole host of ministries:[11] setting up schools for children, sick ministries, medical care, food and clothing distribution, ministering to unwed and destitute mothers, the Stranger's Friend Society (a charity for non-Methodists), ministering to paupers in London, establishing a home for widows in London, establishing an orphanage in Newcastle, unemployment relief, small business loan funding, and prison ministries.[12] This carefully organized world of constraints resulted in sustained accountability and formation of virtue—that is, of moral practices that lead to habit formation. Wesley introduced numerous measures to insure that every member of his classes and bands were carefully looked after. Consequently, the group members that composed the early Methodists tended to overcome negative biological constraints, adopting in their stead altruistic and holy tendencies. His unique approach trained *inward* holiness that led to *outward* actions. We can see contemporary purposeful communities doing the same thing: Reba Place Fellowship in Chicago, Englewood Church in Indianapolis, the Simple Way in Philadelphia, and countless local church groups patterned off of the same accountability model.

NURTURING NATURAL VIRTUES

The best way to have a healthy community that recognizes us as whole people—with our biological roots and environmental influences—is by nurturing the virtues that are naturally with us. If you recall the

---

[11]Stan Ingersol and Wesley Tracy, *Here We Stand: Where Nazarenes Fit in the Religious Marketplace* (Kansas City, MO: Beacon Hill Press, 1999), 10.
[12]For this list, I am indebted to Darrell Moore, "Classical Wesleyanism" (unpublished work, 2011), 6.

argument in previous chapters, we discussed how humans are a genetically mixed bag; we have tendencies that push us to desire vices detrimental to our holiness: lust, gluttony, pride, and the like. But we also possess through evolution naturally developed virtues—qualities like kindness, cooperation, altruism, generosity, and self-sacrifice. It is our task to cultivate these positive virtues while mitigating the vices that draw us away from holy living. This cultivation is no easy task, to be sure, but it is certainly possible. Think back on a time in your life when you realized that a strength of yours could be utilized and nurtured for the good of both yourself and your community.

The life of Amy Carmichael, a missionary to India at the beginning of the twentieth century, is a model of how community can nurture natural virtues to encourage holy behavior. Amy's parents, devout Presbyterians, encouraged her to work alongside them in the Welcome Evangelical Church in Belfast to care for women in the workplace—often poor and downtrodden in that society. This community engaged in purposeful activities, much the same as Wesley's regimented groups. After several years in this purposeful environment, Amy felt a call to serve as a missionary in India, and, despite suffering from a neurological condition that caused chronic pain, Amy embedded herself in a poor Indian community.

While in India, her character—shaped by the methodical virtue formation of her community—was put to the test when she confronted the practice of forced temple prostitution. During this time period, girls would be sold to Hindu temples and married to the deity; such an unholy marriage led to priests and wealthy patrons having sex with the girls so as to be united to the deity. These girls were typically sold at the age of twelve, but often much younger. The patron who paid the parents for the girl would have the privilege of being the first to have sex with her. If the girl complained or escaped, she would be chased down, beaten, and invariably returned to work in the temple.

Amy first found out about this practice when Preena, a little temple girl, fled the temple at night while the guards were sleeping, barely escaping with her life. This flight was the second time Preena had escaped. The first time she ran away from the temple she went to her mother's home, and, for fear of the gods' retribution, her own mother sent Preena back to the temple, where she was tortured with a hot iron. Preena was only seven years old. The next time Preena escaped, she somehow found her way to Amy Carmichael's house and, throwing herself in Amy's arms, begged for help. After explaining to Amy the temple atrocities that happened on a regular basis, Amy was overwhelmed with compassion, perseverance, and generosity—all virtues she had cultivated in her small community of believers. When mobs came to Amy's house and demanded that Amy hand Preena over, Amy didn't move and refused entrance into her home, generously giving Preena the first safe haven she had ever experienced in her life.

Other destitute girls and orphans took notice. In a short time she developed one of the first safe houses in India, where sexually assaulted and abandoned children who were left to exposure could find refuge. Amy became a mother to these children in the orphanage, and many came to call her *Amma* (mother). She personally cared with patient, tender generosity—going against her instincts of self-preservation. Her orphanage cared for over one thousand children and is still in operation today, mostly run by her "children"—those who were nurtured by Amy and shown the patient, courageous way of virtue. Carmichael herself was aware of the great responsibility she had to these children, and she understood that this responsibility called her to give wholly of herself, a notion contrary to self-preservation. Amy saw this selfless love modeled by not only her parents but also her Christian community, a community that practiced purposeful virtue-forming habits.

I realize that missionary stories like this one can often seem so distant and saintly. Yet it is instructive nonetheless because, while it is a spectacular Christian story about virtue, it didn't happen by accident. Amy Carmichael didn't *happen* to be courageous; she cultivated courage over many years in a purposeful community. She didn't *happen* to be generous; she took painstaking time—often to her own detriment—to nurture natural proclivities of altruism while purposefully overcoming urges from her nature to be selfish. Although she seems otherworldly, it's important to remember that while all people who develop the virtues and overcome natural constraints are not saints, all saints like her have developed the virtues and overcome behavioral constraints. And it starts small, by cultivating virtues of kindness, generosity, and selflessness in our own everyday lives.

The Christian church, regardless of denomination, is uniquely positioned to foster virtues through community. Most churches have some kind of small group or accountability structure in place designed to shape those committed members into people who consider how Scripture and holy living form their lives. It seems, then, that these established groups offer us in the church the opportunity to do the sort of work Wesley and Carmichael call us to. And what if these groups considered those areas in our lives that are negatively and positively impacted by our evolutionary roots? What tendencies do we have that hearken back to older parts of the brain, what of our genetic history? Do I come from a family of alcoholics? If that is the case, then I want to be keenly aware of my surroundings and put accountability structures in place to keep me steadfast around alcohol since I'm genetically more prone to this disease. Are you naturally a kindhearted individual? If so, how can you use those natural tendencies to influence those in your group who aren't as biologically or environmentally disposed as you are? How can you cultivate this virtue to sharpen it, deepen it, and make it habit forming? The first step in any journey toward a deeper Christian life is to acknowledge where we are, and this includes where we come from.

But as I extol community groups, I must also add this caveat: I'm not describing just any community group. We have to have the right kind of community, one that purposefully nurtures natural inclinations and natural virtues and works to correct those that lead to selfishness and self-preservation. We've all been part of the wrong kind of social communities—I know I have. Many of us have stories that demonstrate the potential for negative behavior formation through social communities. It was just such a community group that led me as a middle schooler to a friend's house where I smoked my first cigarette. If virtue habit formation can lead us toward a lifestyle of holiness, then we must be intentional about the type of community group we participate in: it has to be the right kind of community.

OLD HABITS DIE HARD

Some ecologists call it the Great Cut of Michigan. During the early nineteenth century, loggers entered the territory of Michigan and began felling trees with axes. From the lumberjacks' estimation, they could cut all the trees in Michigan by slowly working in a northerly direction. By the time they reached Canada, they surmised, the southern trees would have regrown—a natural, sustainable harvest that would support recutting. But technology changed faster than their plans had anticipated, and when in the late 1800s the innovative crosscut saw was brought to Michigan, the lumberjacks began to cut down trees at an exponential rate. So profound was the work of the crosscut saw that in a matter of a few decades most of the trees in Michigan—not to mention the wildlife—were gone. Their old habits did not adjust with the times.

In the same way that the loggers did not learn to adjust, we might take warning as well. While our biological past might have made it advantageous for us to have an unfettered libido and unchecked food consumption, especially with high infant mortality and scarcity of food, contemporary humans are in a vastly different context than we were a thousand millennia ago. We are in a place where our setting has changed and our habits need to be recalibrated. Fortunately, Christians not only have the spiritual tools in place—God's incarnation and the Holy Spirit, the Counselor who is ever with us and convicting us—but also the structural system in place via already existing Christian communities. While we are wedged between our genetic proclivities and their environmental constraints from all sides and in different ways, we have the extraordinary ability to shift how those influences sway our behavior. Close community, as well as personal devotional habits, can propel this movement away from selfish proclivities and toward holy action.

It is imperative that Christians do not expect to live a life of little action, passively letting the grace of God force us into a life of holiness. Nor can we expect to obtain holy living solely according to our own free will. Instead, we must expect to wax and wane with holy living, the motions of the mysterious interplay between free will, constraints on individuals—both biological and environmental—and free grace from God.

For these reasons the Christian individual wishing to develop a life of holiness that produces the fruit of virtuous action must reside within a community where a system of constraints encourages such behavior. Communities that have purposeful structure provide conceptions of the human good to be pursued through the particular virtues that they uphold as central to their notion of what is good.[13] These structures are not necessarily burdensome, but rather they provide a conception of the human good for those who participate in the defining practices of its way of life.[14]

WHAT IS AT STAKE?

Being a Christian in the contemporary world is a high-stakes endeavor. I've never felt that more than when my kids started to get a bit

---

[13]MacIntyre, *After Virtue*, 58.

[14]MacIntyre, *After Virtue*, 58. In a similar way, regarding the connection between evolutionary theory and Christian community, Stephen Pope says, "Christian narrative and participation within the Christian community supply a context for interpreting human evolution, yet the latter gave rise to capacities and inclinations developed within the Christian life. There is a kind of circulatory, but not a vicious one, in the relation between faith and nature in this regard" (Pope, *Human Evolution and Christian Ethics*, 296).

older. My ten year old is starting to ask difficult questions that actually matter. Up to this point in his life, his questions were almost always concerned with how much candy he could have. Now, he's starting to ask questions about why humans are on the planet and where we came from. He's also asking more mature questions about the birds and the bees. Though it's only natural for him to do so, it's likewise natural for me to be terrified about the conversation. My wife and I have decided not to sugarcoat reality for our children. And, for the same reasons that we tell them that Santa Claus isn't real, we're honest about our evolutionary roots and our dependence on tradition to help us understand the Bible.

When my oldest asked how we were related to primates, I didn't shy away from the conversation. It was remarkable how quickly he accepted the idea of common ancestry and then made this astounding observation: "I guess that makes all of creation pretty special, huh, Dad?" Yes, it does! We don't have to be afraid of our common ancestry; we can celebrate it. In fact, the incarnation is probably the best place to start looking at how all of creation is intimately connected and exceptional. God almighty became incarnate as a human person, which makes us very special. Moreover, because humans have a common ancestry with complex animals, that means that Christ didn't just become a human, he became an animal—making all animals special, as God also became one of them. Furthermore, humans are not just complex animals— we're organic, making Christ's incarnation important for all organisms. Even greater still, humans are material—we have material bodies that cannot be denied—so Christ's incarnation didn't just penetrate the organic world, it became infused into the material world, making all of creation truly special. Indeed, humans are distinct because of the *imago Dei*; we cannot ignore the fact that Jesus came down not just as any animal but as a human animal. But it is likewise true that, in light of our common ancestry, the incarnation elevates all things, not just us. To me, that's a hopeful and inspiring message.

In this book we've discussed the necessity of reading Scripture through the lens of our tradition and the early church fathers; we didn't shy away from exploring some of the scarier sides of evolutionary theory for the Christian faith, and we looked at how embracing evolutionary theory can actually help our personal journeys toward holiness by helping us not only to be aware of our roots but also to understand how to nurture natural virtues and overcome natural vices. Yet it is important to keep in mind that we are to be truthful to the next generation of Christians. To look the other way and to keep our evolutionary roots at arm's length is not only untruthful to our fellow believers, but it is also, as Augustine pointed out in the fourth century, detrimental to our witness.[15] Embracing evolution should not be misconstrued as a mere acquiesce to science but rather seen as a potential boon to our journeys toward holy living.

---

[15] Augustine, *The Literal Meaning of Genesis*.

STUDY GUIDE

CHAPTER 1: OPENING A DIALOGUE
1. In what ways are you excited or nervous about opening up a dialogue between evolution and your Christian faith?
2. What role has science played in your life as a Christian in the church?
3. Why do you think science is seen by some in the church as antithetical to the faith?
4. What would it look like if you allowed yourself to be open to an evolutionary creation perspective? How would it affect your faith?

CHAPTER 2: READING SCRIPTURE FAITHFULLY
1. How have your personal experiences established the lens that you use to view Scripture?
2. How has your experience in the church affected the lens you use to view the topic of evolution?
3. Did anything in this chapter challenge what you've previously thought, believed, or been taught? In what ways?
4. If you were to adopt an integrative approach to science and Christianity, what would you have to give up? What would you gain?

CHAPTER 3: ADAM AND EVE, THE FALL, PREDATION, AND DEATH
1. In what ways have you understood earthly suffering and death in relationship to Christianity?
2. How might a partnership with science instead of a polarization help in understanding the problem of death and suffering?
3. This chapter explored the idea that there has to be death in order for progression and new life to happen. How can understanding this concept help us in our faith?
4. Are there any other troubling theological implications that come about by adopting an integrated perspective of science and Christianity?

CHAPTER 4: THE NUTS AND BOLTS OF EVOLUTION
1. How did reading about the nuts and bolts of evolution make you feel about evolutionary creationism? Why do you think that is?
2. In what ways does understanding our evolutionary roots make you see the world differently?
3. How can this understanding shape your sense of connection to the organic world around you?
4. How is God calling you to take care of the world and his creation?

CHAPTER 5: RELATING TO SCIENCE

1. How can you look at science holistically—recognizing the good that it gives us and what it's telling us about our past?
2. What experiences have you had with the evolution-versus-creationism debate, and how have these experiences impacted your reading of this book?
3. In what ways do you see our evolutionary roots being harmonious with our Christian faith?
4. In what ways do you struggle to fit evolution and Christianity together?

CHAPTER 6: UNDERSTANDING EVOLUTIONARY THEORY CAN BE EMPOWERING

1. How does belief in evolution deepen or change the way you see your relationship to God or God's relationship to creation?
2. In what ways can evolutionary theory help make sense of our own human nature? In what practical ways can we use this knowledge?
3. How are some areas in your past—regrettable actions or sinful proclivities—better understood by looking at them through a spiritual *and* evolutionary lens?
4. In what ways can you communicate these concepts to the next generation in order to help better protect them from negative biological proclivities?

CHAPTER 7: HAVING EVOLUTIONARY ROOTS ISN'T JUST BAGGAGE

1. How does understanding our evolutionary tendencies change our theology of sin and personal responsibility to control our actions?
2. In what ways can you harness your positive and altruistic evolutionary roots in order to become a healthier Christian?
3. How can you nurture natural tendencies to reach your neighbors for the kingdom?
4. In what ways can you encourage children in their developmental stages to cultivate positive behavioral traits?

CHAPTER 8: NURTURING NATURAL VIRTUES TOWARD HEALTHY COMMUNITY

1. In what ways have you experienced unhealthy Christian community (compared to the healthy type of community discussed at the end of this chapter)?
2. How does your community inhibit or encourage spiritual growth?
3. What practical parameters can you put in your life in order to nurture healthy community? In other words, what can you *actually do* to develop a healthy Christian life?
4. What virtues does every healthy Christian community need to have?

BIBLIOGRAPHY

Applegate, Kathryn, and J. B. Stump, eds. *How I Changed My Mind About Evolution: Evangelicals Reflect on Faith and Science*. Downers Grove, IL: IVP Academic, 2016.

Augustine. *The Literal Meaning of Genesis*. Translated and edited by John Hammond Taylor. Mahwah, NJ: Paulist Press, 1982.

Bangs, Carl. *Our Roots of Belief: Bible Faith and Faithful Theology*. Kansas City, MO: Beacon Hill Press, 1981.

Barbour, Ian G. *Issues in Science and Religion*. Englewood Cliffs, NJ: Prentice-Hall, 1966.

Barrett, Deirdre. *Supernormal Stimuli: How Primal Urges Overran Their Evolutionary Purpose*. New York: W. W. Norton, 2010.

———. *Waistland: A (R)Evolutionary View of Our Weight and Fitness Crisis*. New York: W. W. Norton, 2007.

Berry, Wendell. *The Country of Marriage*. Berkeley, CA: Counterpoint, 2013.

———. *Life Is a Miracle: An Essay Against Modern Superstition*. Washington, DC: Counterpoint, 2000.

Boyd, Craig A. "Thomistic Natural Law and the Limits of Evolutionary Psychology." In *Evolution and Ethics: Human Morality in Biological and Religious Perspective*, edited by Philip Clayton and Jeffrey Schloss, 221-37. Grand Rapids: Eerdmans, 2004.

Carballo, David M. *Cooperation & Collective Action Archaeological Perspectives*. Boulder, CO: University Press of Colorado, 2013.

Clayton, Philip. "Biology and Purpose: Altruism, Morality, and Human Nature in Evolutionary Perspective." In *Evolution and Ethics: Human Morality in Biological and Religious Perspective*, edited by Philip Clayton and Jeffrey Schloss, 318-39. Grand Rapids: Eerdmans, 2004.

———. "Evolution, Altruism, and God: Why the Levels of Emergent Complexity Matter." In *Evolution, Games, and God: The Principle of Cooperation*, edited by Martin A. Nowak and Sarah Coakley, 337-400. Cambridge, MA: Harvard University Press, 2013.

Cunningham, Conor. *Darwin's Pious Idea: Why the Ultra-Darwinists and Creationists Both Get It Wrong*. Grand Rapids: Eerdmans, 2010.

de Waal, Frans. *The Age of Empathy: Nature's Lessons for a Kinder Society*. New York: Harmony Books, 2009.

Dennett, D. C. *Darwin's Dangerous Idea: Evolution and the Meanings of Life*. New York: Simon & Schuster, 1995.

Dowd, Michael. *Thank God for Evolution! How the Marriage of Science and Religion Will Transform Your Life and Our World*. Tulsa, OK: Council Oak Books, 2007.

———. "Why We Struggle Now." TEDx, Grand Rapids, June 14, 2012, www.youtube.com/watch?v=DDMOF7qtlh8.

Ecklund, Elaine Howard, and Christopher P. Scheitle. *Religion vs. Science: What Religious People Really Think*. New York: Oxford University Press, 2017.

Falk, Darrel R. *Coming to Peace with Science: Bridging the Worlds Between Faith and Biology*. Downers Grove, IL: InterVarsity Press, 2004.

———. "Human Origins: The Scientific Story." In *Evolution and the Fall*, edited by William T. Cavanaugh and James K. A. Smith, chap. 1. Grand Rapids: Eerdmans, 2017.

Finocchiaro, Maurice A. *The Galileo Affair: A Documentary History*. California Studies in the History of Science. Berkeley: University of California, 1989.

Fugle, Gary N. *Laying Down Arms to Heal the Creation-Evolution Divide*. Eugene, OR: Wipf & Stock, 2015.

Galilei, Galileo. "Letter to the Grand Duchess Christina of Tuscany." 1615.

Giberson, Karl, and Francis S. Collins. *The Language of Science and Faith: Straight Answers to Genuine Questions*. Downers Grove, IL: IVP Books, 2011.

Gould, Stephen Jay. *Punctuated Equilibrium*. Cambridge, MA: Belknap Press, 2007.

Green, Joel B. *Body, Soul, and Human Life: The Nature of Humanity in the Bible*. Studies in Theological Interpretation. Grand Rapids: Baker Academic, 2008.

———. *What About the Soul? Neuroscience and Christian Anthropology*. Nashville: Abingdon Press, 2004.

Gregory of Nyssa. *The Life of Moses*. Classics of Western Spirituality. New York: Paulist Press, 1978.

Hegel, Georg Wilhelm Friedrich. *Lectures on the Philosophy of Religion*. Translated by P. C. Hodgson R. F. Brown, J. M. Stewart. Edited by Peter C. Hodgson. One-volume ed. Berkeley: University of California Press, 1988.

Hill, Matthew Nelson. *Evolution and Holiness: Sociobiology, Altruism, and the Quest for Wesleyan Perfection*. Strategic Initiatives in Evangelical Theology. Downers Grove, IL: IVP Academic, 2016.

Holsinger-Friesen, Thomas. *Irenaeus and Genesis: A Study of Competition in Early Christian Hermeneutics*. University Park, PA: Eisenbrauns, 2009.

Ingersol, Stan, and Wesley Tracy. *Here We Stand: Where Nazarenes Fit in the Religious Marketplace*. Kansas City, MO: Beacon Hill Press, 1999.

Küng, Hans. *Theology for the Third Millennium: An Ecumenical View*. New York: Doubleday, 1988.

Lubac, Henri de. *Medieval Exegesis: The Four Senses of Scripture*. Ressourcement. Grand Rapids: Eerdmans, 1998.

MacIntyre, Alasdair C. *After Virtue: A Study in Moral Theory*. 3rd ed. Notre Dame, IN: University of Notre Dame Press, 2007.

Maddox, Randy L. *Responsible Grace: John Wesley's Practical Theology*. Nashville: Kingswood Books, 1994.

McGrath, Alister E. *Science and Religion: An Introduction*. 2nd ed. Malden, MA: Wiley-Blackwell, 2010.

McHenry, Henry M. "Human Evolution." In *Evolution: The First Four Billion Years*, edited by Michael Ruse and Joseph Travis, 961-85. Cambridge, MA: Belknap Press, 2009.

Moore, Darrell. "Classical Wesleyanism." Unpublished work, 2011.

Mulholland, M. Robert. *Shaped by the Word: The Power of Scripture in Spiritual Formation*. Rev. ed. Nashville: Upper Room Books, 2000.

Nowak, Martin A. "Five Rules for the Evolution of Cooperation." In *Evolution, Games, and God: The Principle of Cooperation*, edited by Martin A. Nowak and Sarah Coakley, 337-400. Cambridge, MA: Harvard University Press, 2013.

Oden, Thomas C. *John Wesley's Teachings: Christ and Salvation*. Vol. 2. Grand Rapids: Zondervan, 2012.

Patterson, Nick, Daniel J. Richter, Sante Gnerre, Eric S. Lander, and David Reich. "Genetic Evidence for Complex Speciation of Humans and Chimpanzees." *Nature* 441, no. 7097 (June 2006): 1103-8.

Pope, Stephen J. *The Evolution of Altruism and the Ordering of Love*. Moral Traditions and Moral Arguments. Washington, DC: Georgetown University Press, 1994.

BIBLIOGRAPHY

———. *Human Evolution and Christian Ethics*. New Studies in Christian Ethics. Cambridge, MA: Cambridge University Press, 2007.

———. "Relating Self, Others, and Sacrifice in the Ordering of Love." In *Altruism and Altruistic Love: Science, Philosophy, and Religion in Dialogue*, edited by Stephen G. Post, Lynn G. Underwood, Jeffrey P. Schloss, and William B. Hurlbut. New York: Oxford University Press, 2002.

Post, Stephen G., Lynn G. Underwood, Jeffrey P. Schloss, and William B. Hurlbut, eds. *Altruism and Altruistic Love: Science, Philosophy, and Religion in Dialogue*. New York: Oxford University Press, 2002.

Quammen, David. *The Tangled Tree: A Radical New History of Life*. New York: Simon & Schuster, 2018.

Ridley, Matt. *The Red Queen: Sex and the Evolution of Human Nature*. New York: Perennial, 2003.

Rolston, Homes. "The Good Samaritan and His Genes." In *Evolution and Ethics: Human Morality in Biological and Religious Perspective*, edited by Philip Clayton and Jeffrey Schloss, 318-39. Grand Rapids: Eerdmans, 2004.

Ruse, Michael, and Joseph Travis, eds. *Evolution: The First Four Billion Years*. Cambridge, MA: Belknap Press, 2009.

Sober, Elliott, and David Sloan Wilson. *Unto Others: The Evolution and Psychology of Unselfish Behavior*. Cambridge, MA: Harvard University Press, 1998.

Swamidass, S. Joshua. "The Overlooked Science of Genealogical Ancestry." *Perspectives on Science and Christian Faith* 70, no. 1 (March 2018): 19-35.

Wagner, Andreas. *Arrival of the Fittest: Solving Evolution's Greatest Puzzle*. New York: Penguin Group, 2014.

Walton, John H. *The Lost World of Adam and Eve: Genesis 2-3 and the Human Origins Debate*. Downers Grove, IL: InterVarsity Press, 2015.

Weil, Simone. *Waiting for God*. New York: Putnam, 1951.

Weiss, Kenneth M., and Anne V. Buchanan. *The Mermaid's Tale: Four Billion Years of Cooperation in the Making of Living Things*. Cambridge, MA: Harvard University Press, 2009.

Wesley, John, *John Wesley*. Edited by A. C. Outler. New York: Oxford University Press, 1980.

———. *Sermons 71-114*. Edited by Albert C. Outler. The Bicentennial Edition of the Works of John Wesley. Nashville: Abingdon Press, 1986.

———. *The Works of John Wesley*. Vol. 8. Edited by Thomas Jackson and Albert C. Outler. Grand Rapids: Zondervan, 1958.

Wesley, John, and Rupert E. Davies. *The Methodist Societies: History, Nature, and Design*. The Bicentennial Edition of the Works of John Wesley. Nashville: Abingdon Press, 1989.

Wesley, John, and Charles Wesley. *Hymns and Sacred Poems*. Bristol, UK: F. Farley, 1739.

Wilson, David Sloan. *Evolution for Everyone: How Darwin's Theory Can Change the Way We Think About Our Lives*. New York: Delta Press, 2007.

Wilson, Edward O. *The Social Conquest of Earth*. New York: Liveright, 2012.

Wright, N. T. *Surprised by Hope: Rethinking Heaven, the Resurrection, and the Mission of the Church*. New York: HarperOne, 2008.

Wynkoop, Mildred Bangs. *A Theology of Love: The Dynamic of Wesleyanism*. Kansas City, MO: Beacon Hill Press, 1972.

Finding the Textbook You Need

The IVP Academic Textbook Selector
is an online tool for instantly finding the IVP books
suitable for over 250 courses across 24 disciplines.

ivpacademic.com

Printed in the USA
CPSIA information can be obtained
at www.ICGtesting.com
CBHW030931221123
2041CB00031BA/145